REPAIRING PARADISE

NPS PHOTO BY JIM PEACO

REPAIRING PARADISE

THE RESTORATION OF NATURE IN AMERICA'S NATIONAL PARKS

WILLIAM R. LOWRY

BROOKINGS INSTITUTION PRESS
Washington, D.C.

THE BROOKINGS INSTITUTION
1775 Massachusetts Avenue, N.W., Washington, D.C. 20036
www.brookings.edu

The Library of Congress has cataloged the hardcover edition as folllows:
Lowry, William R. (William Robert), 1953
Repairing paradise: the restoration of nature in America's national parks / William R. Lowry.
p. cm.
Includes bibliographical references and index.
Summary: "Examines whether the U.S. can restore the most-loved crown jewels of its national park system, focusing on four ambitious efforts to reverse environmental damage. Combines field research with public policy analysis to portray the mission to restore the natural health and glory to some of the world's most wondrous places"—Provided by publisher.
ISBN 978-0-8157-0274-0 (hardcover : alk. paper)
1. National parks and reserves—Government policy—United States. 2. National parks and reserves—United States—Management. 3. National parks and reserves—Environmental aspects—United States. 4. Conservation of natural resources—United States. 5. Nature conservation—United States. 6. Environmental policy—United States. 7. United States—Environmental conditions. I. Title.
SB482.U6L68 2009
363.6'80973—dc22 2009027073
ISBN 978-0-8157-2270-0 (pbk : alk. paper)

First paperback edition January 2012

Digital printing

Printed on acid-free paper

All photographs by the author except as noted

Typeset in Sabon

Composition by Cynthia Stock
Silver Spring, Maryland

For Lynn, Joey, and Tom

CONTENTS

ACKNOWLEDGMENTS

As will quickly become apparent to anyone reading my personal anecdotes in this book, researching and producing it has been a labor of love. Many people helped at various stages, but the only thanks I can offer them here is a brief acknowledgment. The field work was (no surprise) especially enjoyable. Those who helped to make it so include the Bennetts, the Blacks, the Browns, the Brunis, Gary Ferguson and Martha Young, the Gustasons, Steve Gehman and Betsy Robinson, the Ladds, Pam Lokken and Andy Sobel, the Opekas, and, of course, the Lowry family. As I mention in the book, I appreciate the time and insights given to me by the people involved in the efforts to repair the parks. While I can't name all of them here, several warrant special thanks, notably Doug Smith and Mike Yochim at Yellowstone, Linda Dahl at Yosemite, Stuart Appelbaum and Jeff Jacobs at Everglades, and Jan Balsom and Sam Spiller at Grand Canyon. Colleagues here at Washington University who helped include Bill Bottom, Joe Frank, Melinda Kramer, Andy Mertha, Gary Miller, Ian Ostrander, Itai Sened, Steve Smith, and Andy Sobel. Thanks to Chris Kelaher, Bob Faherty, Janet Walker, Eileen Hughes, Larry Converse, two anonymous referees, and the Brookings staff who shaped this final product. At various times during this project, I used and appreciated funding from the Council for International Exchange of Fulbright Scholars, the Washington University Graduate School of Arts and Sciences, and the Weidenbaum Center. If I've forgotten someone who deserves mention, I apologize. Thanks for your help.

Finally, this book is dedicated to Lynn, Joey, and Tom. Much as I love being in the parks, I always like to come home, and that's above all because of them.

REPAIRING PARADISE

Half Dome with Vernal and Nevada Falls, viewed from Glacier Point in Yosemite National Park.

1
CHANGING POLICIES IN NATIONAL PARKS

It's time for us to change America.

—Barack Obama

If you want to make enemies, try to change something.

—Woodrow Wilson

The overarching theme of U.S. politics today is change. As suggested by the opening quote, which comes from Barack Obama's speech accepting the Democratic Party's nomination for president, Obama built his campaign on that theme. Not to be outdone, his opponent, John McCain, countered Obama's slogan, "Change you can believe in," with "Change you can trust."[1] But as the famous quote from former President Wilson warns—and as President Obama is now fully aware—significant change does not come easily. How, then, does it occur?

Natural resource management is one area in which policymakers have sought to make significant changes. In the closing decades of the twentieth century, many Americans realized that traditional natural resource policies had resulted in substantial and often negative impacts on the environment. Whether intentionally or not, those policies had diminished or destroyed or at least altered many of the most precious public lands and rivers on the North American continent. Some Americans began to think about taking the next step in their evolving relationship with nature, an evolution that historically has witnessed stages of fear, ignorance, abuse, use, and finally preservation. But preservation was not enough. Conditions had been changed, so simply saving public sites as they now were might preserve what many viewed as the mistakes of the past. The next step involved restoration—or if that was not possible, then at least repair of those mistakes. Thus, a decade into the twenty-first century, decisionmakers have promised to change policies to restore natural conditions at literally hundreds of public sites across the United States.

This book discusses efforts to change traditional policies at four sites that are important to millions of people, not just U.S. citizens. The names of the places alone—Yellowstone, Yosemite, the Everglades, and the Grand Canyon—inspire images of deep forests, majestic peaks, extraordinary wildlife, and vistas that can overwhelm anyone's senses. To a considerable extent, those images are still accurate, but public policies have caused significant deterioration of natural conditions at these magnificent places. Although the mandate for the National Park Service, the public agency responsible for national parks, calls on park officials to leave them "unimpaired" for future generations, the parks already have been impaired. In recent years, therefore, policymakers have attempted to reverse traditional policies by reintroducing eliminated species, reducing automobile traffic, replenishing fresh water supplies, and restoring natural water flows. Are the goals of those efforts being realized? I will argue that those seeking to alter the status quo can achieve the substantial change necessary to repair the damage from entrenched, traditional practices only when they create effective coalitions for change.

The Challenge of Changing Past Policies

Before addressing how such repairs might be made, I emphasize that democratic political systems, certainly including the U.S. government, generally are not conducive to effecting dramatic change. Indeed, the framers designed the U.S. government to include separation of powers and multiple checks and balances in order to make radical change difficult. An obvious example is the procedure required to amend the Constitution, a procedure that has proven to be an insurmountable obstacle for causes ranging from equal rights for women to prohibition of flag burning. Even beyond the institutional obstacles to change, policies themselves, once implemented, take on a life of their own that is not easy to alter. Procedures become routine, and interests of all kinds become entrenched. One of the most powerful concepts in the policy literature is that of the "iron triangle" made up of interest groups, public agencies, and members of Congress. Such coalitions have come to dominate specific policy issues, working together to benefit from the status quo.[2] Not surprisingly then, the first principle taught in many courses on public policy is that change typically occurs only incrementally or inconsistently.[3] Even prominent scholars who have questioned that principle acknowledge that incrementalism "has dominated thinking about policy change since the 1950s."[4]

However difficult substantial change may be, policymakers do occasionally pronounce some event as signaling an entirely new policy approach. In fact, some public policies have changed so dramatically over time that they are considered reversals of past goals. Whereas for decades state and local governments pursued policies designed to prevent the participation of black Americans in the political process, for example, federal mandates in 1964 and 1965 not only declared such policies illegitimate but implemented others, such as the designation of minority congressional districts, to reverse those policies. Policies on other issues, from pesticide regulation to discouragement of smoking to decommissioning of nuclear power plants, have displayed dramatic changes over time.[5] The frequency of such changes, notably in policy areas involving environmental issues, seems to have increased in recent decades, perhaps due to improvements in science, greater understanding, better communication, the immediacy of events, or even the willingness to question any traditional behavior. However, efforts to achieve significant changes may be even more daunting in coming years if the country faces continued partisan gridlock and calls for increased fiscal austerity measures.

Restoring natural environments requires substantial changes in policy goals and human behavior. Many scientists argue that restoring an ecosystem to its condition prior to human disturbance is virtually impossible if for no other reason than the fact that environments are always changing,[6] and many restoration proponents realize that achieving an exact replica of an earlier ecosystem may not be possible. A more operational definition, then, is "returning an ecosystem to a close approximation of its condition prior to disturbance."[7] The Society for Ecological Restoration identifies restoration as "reestablishing an ecologically healthy relationship between nature and culture."[8]

To assess the efforts to repair the damage to natural conditions from past policies, this analysis focuses on projects at four of the most revered national parks: Yellowstone, Yosemite, the Everglades, and the Grand Canyon. The World Heritage Committee lists all four as World Heritage sites, along with such remarkable places as the Great Barrier Reef and the Galapagos Islands. The outcomes of repair efforts at these parks are important for at least two reasons. First, as anyone who has ever visited them knows, the reverence accorded to these places is well deserved. Thus, what happens to them is of interest to the millions of people who value public lands and waters. Second, the outcome of restoration efforts here will say much about restoration efforts elsewhere. To put it more bluntly, if repair efforts here are doomed to failure, then what can be expected of restoration projects at other, less revered sites?

The Tarnished Crown Jewels

However destructive the policies that followed, the basic decision to set aside some of the nation's most precious lands and waters in a national park system showed remarkable foresight. One can argue whether statutory protection of lands in the mid-nineteenth century resulted from the efforts of prescient conservationists or tourist-seeking entrepreneurs.[9] The fact is that federal policymakers, beginning with the original assignment of Yosemite to the state of California in 1864 and the designation of Yellowstone as a national park in 1872, initiated a process that resulted in protection of a truly remarkable set of public spaces. Today, the national park system contains nearly 400 of the most scenic, historic, and valued sites in the country. While these sites still entice and inspire millions of visitors each year, as shown in figure 1-1, there's trouble in paradise.

Figure 1-1. *National Parks Visitation, 1900–2007*

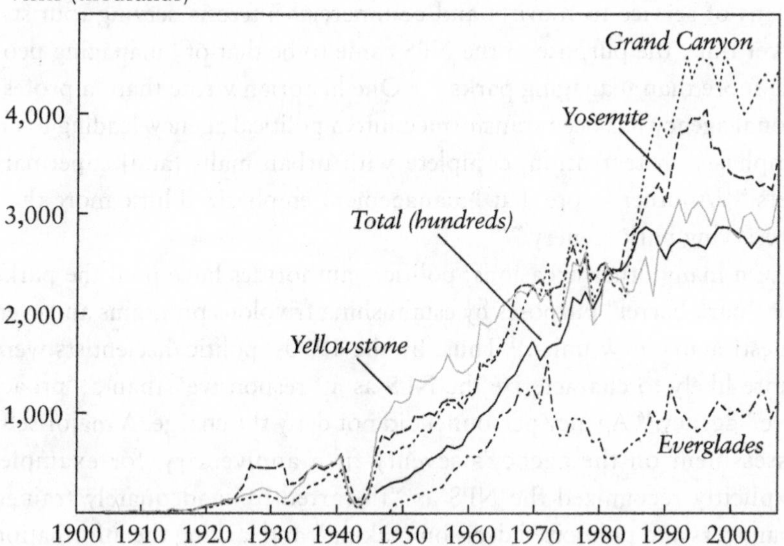

Background

As the park system grew, in 1916 Congress established the National Park Service (NPS) within the Department of the Interior to manage the system's various units, and in so doing it created a lasting source of tension in park operations. Congress mandated the NPS to facilitate "enjoyment" of the parks even while keeping them in "unimpaired" condition for generations to come. The NPS has since struggled to find a balance between those potentially competing missions, a balance that may never have existed in the first place. To a significant extent, this struggle created a legacy for the NPS that affects its implementation of any program, including the most high-profile.

The mandate to provide both use and preservation has created an agency with something of a split personality. Most within the NPS are determined to emphasize preservation, often joking of being "paid in sunsets." One study found agency employees bonded by a "deep faith in the idea of a national park system."[10] A survey of NPS personnel in the 1980s found 84 percent seeing preservation as "the major purpose" of the agency while only 9 percent emphasized use.[11] However, political authorities have often pushed the NPS to promote use, especially in terms of service to tourists and commercial interests serving tourists. Over time, the purpose of the NPS came to be that of "managing people more than managing parks."[12] One historian wrote that "a professional agency has been transformed into a political agency, leading to an emphasis on recreation, complete with urban malls [and] supermarkets."[13]Another wrote that "management emphasized little more than preserving park scenery."[14]

On many other occasions, political authorities have used the parks for "park barrel" purposes by establishing frivolous programs and even questionable new units.[15] Thus, by the 1990s, political scientists were more likely to characterize the NPS as a "responsive" than a "proactive" agency.[16] Agency personnel did not deny the charge. A major self-assessment on the agency's seventy-fifth anniversary, for example, explicitly recognized the NPS as "thwarted by inadequately trained managers and politicized decision making" and lacking the information and capability needed to "defend its mission and resources in Washington."[17] As a result, agency personnel are generally sympathetic to the goals of restoration and preservation but somewhat risk-averse in implementing the changes to achieve them.

NPS personnel also face significant challenges in their daily management of the parks. The parks face both internal threats (such as traffic and congestion) and external threats (such as pollution and encroaching development). The most obvious internal threat involves excessive visitor use. The line showing total visitation to national parks in figure 1-1 reflects the success of the NPS in attracting increasing numbers of visitors throughout the twentieth century. Visits are shown in thousands. In order to put the figure on the same scale as the individual park graphs, the line for total visits actually reflects total visits divided by 100, but the overall trends are telling. The problems arise not from the number of people coming into the parks as much as from how they get there and how they use them once inside. (The leveling off in the last two decades, discussed in later chapters, is a concern for the agency.)

As for external threats, the parks proved disturbingly vulnerable. The previously mentioned self-assessment described "a mismatch between the demand that the park units be protected and the tools available when the threats to park resources and values are increasingly coming from outside unit boundaries."[18] Those threats include air and water pollution from outside sources as well as encroachment by growth and development outside park boundaries. Management reactions to those threats often were ineffective and contributed to the problems in the national parks. At some places, including those discussed in this book, the consequences have been substantial.

The result of the agency's dual mandate, political demands, and the daunting threats to the parks often were policies that did not ensure preservation of the natural environment and frequently caused substantial damage instead.

A New Era?

The environmental awakening of the late twentieth century led to potentially significant changes in the management of the national parks. In short, as one analyst stated, people increasingly recognized that "human attempts to dominate nature could well have a dark underside."[19] One aspect of those changes involved renewed attention to the goal of protecting the natural conditions within the units. For example, the first strategic objective listed by the seventy-fifth anniversary report was that "the primary responsibility of the NPS must be protection of park resources from internal and external impairment."[20] A second aspect of those changes was more aggressive. Policymakers issued new orders for management not just to protect or even preserve parks but to work to restore natural conditions within parks that had suffered from past practices.

At Yellowstone, traditional wildlife policies had had a severe impact on the ecosystem because they removed predators, but in the 1990s, NPS and the Fish and Wildlife Service (FWS) reintroduced wolves to the park, attempting to restore a balance that had been absent since the 1930s. Yosemite's managers historically allowed and in fact encouraged the use of automobiles to visit and explore the park, but in recent years policymakers have called for the reduction and eventual elimination of automobile traffic in the heart of the ecosystem, the Yosemite

Valley. This plan, first launched in 1980, was reiterated in somewhat different form in 2000.

Historical policies at the Everglades diverted the water so essential for these wetlands to rapidly growing cities and farms to the point that the ecosystem was nearly destroyed. In 2000, however, policymakers launched the most expensive restoration plan ever attempted, with the stated goal of repairing the damage from past water allocation and use practices. Past policies at the Grand Canyon, in particular the construction and use of a large hydroelectric dam in 1963 at Glen Canyon, substantially altered the ecosystem of the Colorado River. Beginning in the early 1990s, policymakers promised to try to restore some of that ecosystem, while some environmental groups argued for removal of the Glen Canyon Dam.

Why These Cases?

The cases in this book were chosen for several reasons. First, as mentioned, they involve some of the most precious areas in the world. By virtually any standard, the four parks are the cream of the crop, the diamonds among the crown jewels that make up the U.S. national park system. Furthermore, all four are World Heritage sites, renowned not just in the United States but throughout the world. Efforts to restore them therefore are important to literally millions of people. Second, because these are such revered lands, the repair projects are high-profile, and that fact has implications for other restoration efforts. The failure of these efforts could result in severe criticism and negative consequences for funding of similar projects. Explaining failures, or even limited successes, as is done in this book, can mitigate blanket condemnations of efforts to restore natural conditions in times of fiscal scarcity. A third reason is that these cases are quite similar in many ways and therefore comparable. All take place in highly valued natural areas where every action is publicized and of importance at least to certain parts of society. All enjoy considerable support among the environmental community. Restoration projects at all four parks require significant departures from past practices; thus, at least some conflict is inevitable at all four. All four involve implementation of efforts to change policies during the last fifteen years. A fourth reason is that these cases represent a range of challenges involving public lands, from species protection to visitor use to external encroachment to river corridor management. Two represent largely internal issues and two largely external threats. Finally, the cases differ in ways that are important to more general arguments regarding changes to traditional public policies.

A Framework for Assessing Repair Efforts

A large literature addresses broad questions of democratic responsiveness and policy change, illustrating useful concepts for anticipating and assessing the efforts to repair the damage to parks from past policies. Those concepts include the importance of coalitions for change and the conditions that they create and use to gain support for change.

Coalitions for Change

Virtually every effort to change the status quo involves conflict between those who benefit from the status quo and those who are trying to alter it. The reintroduction of wolves to Yellowstone, for instance, has produced conflicts with those who prefer their absence. As two pioneering social scientists stated, "many decision problems take the form of a choice between retaining the status quo and accepting an alternative to it, which is advantageous in some respects and disadvantageous in others."[21] Change does not come automatically, but as the result of the actions of those who seek it. How do they proceed?

Whatever the goal—to bring back wolves to Yellowstone, reduce auto traffic in Yosemite, restore water flows in the Everglades, or remove the Glen Canyon Dam—advocates have to apply pressure on political leaders. A long line of literature on pressure politics in the United States has evolved, providing a useful tool for assessing the actions of advocates for change. The idea that pressure groups formed around common interests in order to affect public policies has been prevalent at least since James Madison's fear of factions in *Federalist 10*. Well into the twentieth century, many political scientists argued that to understand policies, you needed to understand the interest groups trying to affect those policies;[22] some even argued that that was all you needed to understand.[23] Those arguments have evolved over time. One widely adopted perspective in the policy literature on political pressure posits a crucial role for advocacy coalitions. The advocacy coalition framework describes policy as a product of stable system parameters such as constitutional rules, changes in the external environment such as economic crises, and competition between coalitions of actors. This framework has proven a compelling tool in the analysis of dozens of different instances of policy change.[24]

Advocacy coalitions contain not just interest groups but also journalists, researchers, and agency officials who "seek to influence public policy" in a particular domain.[25] Analysis based on this conceptualization is a significant improvement over past analyses that examined only the impact of interest groups. Indeed, what some saw as the diminished influence of individual groups may well have been the result of a misguided effort to assess the influence of just one isolated component of a larger collection of institutional actors all pursuing a similar goal. It also extends the work of scholars who questioned the utility of "iron triangles," with their focus on only three components: interest group leaders, sympathetic agency officials, and complicit legislators.[26]

The idea that advocacy coalitions contain journalists, academics, and agency officials is helpful in defining and analyzing pressure for change, but it is still admittedly somewhat "undeveloped."[27] Advocacy coalition scholars suggest that the different components of a coalition "engage in a nontrivial degree of coordination."[28] I argue that the degree of coordination is less important than the fact that the different parts of even an informal coalition are working toward the same goals. How, in particular, do those different parts help make the overall effort more effective? At each of the parks in question, interest groups, journalists, academics, and agency officials have all played important roles in repair efforts, albeit to varying degrees of effect. I argue that the inclusion of these actors, regardless of how much they explicitly coordinate, helps explain how pro-change forces create and use certain conditions to effectively engage the larger public.

Conditions for Change

If a coalition wants to make a dramatic change, such as reintroducing an eliminated species or removing a dam, what can it do to succeed?

One of the principal lessons from the literature on policy change is that altering the status quo requires the involvement of the larger public. E. E. Schattschneider described any conflict situation as involving both participants and observers and argued, "The first proposition is that the outcome of every conflict is determined by the extent to which the audience becomes involved in it."[29] The audience consists of the many people who have not previously been involved in the conflict but whose collective action can, if mobilized, foster fundamental changes to the status quo. They can provide the pressure, resources, and financial support essential to formulating and implementing significant policy changes. How can coalitions create and expand the sphere of conflict to make audience involvement in repair efforts more likely? I propose several approaches: they can define the issue; present economic arguments; present scientific evidence; and elicit agency commitment.

Issue Definition. First, the audience is more likely to become involved if the image of the alternative to the status quo is more positive than that of the status quo. For instance, the people who seek to eliminate cars from Yosemite Valley face arguments from those who defend the public's right to easy access to the park. Change advocates therefore have to portray their goal as something other than an attempt to "lock people out" of the park—perhaps as an attempt to preserve the natural beauty of the place, which is what drew the public in the first place.

Much of the argument for the importance of issue definition comes from recent work that attempts to explain periods of dramatic change in policies but allows for what scholars term "punctuated equilibrium": policies typically are stable, with only marginal adjustments, except for periods during which something happens to produce dramatic change.[30] The "something" that happens is substantial change in the image of an issue. The policy image involves "the supporting set of ideas structuring how policymakers think about and discuss the policy."[31] For instance, nuclear policy was in relative equilibrium until the image of nuclear power shifted from one that focused on energy generation to one that focused on public safety and environmental damage. Once the image of an issue has been shaken, change to the status quo is much more possible.[32] Scholars have shown the importance to policy changes from image alteration in places ranging from Canada to China.[33]

How can pro-change coalitions define issues to engage the larger public? One way is to frame issues in terms that are sympathetic to proposed changes or antipathetic to the status quo. Framing highlights some element of reality regarding an issue to the point of affecting perceptions of that issue.[34] For example, policies affecting a river that have traditionally been framed in terms of increasing economic utility may be described in terms of decreasing environmental health. Advocates of eliminating inheritance taxes use the term "death taxes" rather than "estate taxes." At Yosemite, those seeking to reduce cars in the valley could frame the status quo in terms of traffic jams and smog. Another way to shape issue definition is to use new information that can "shock, disrupt, and destabilize" a previously stable policy.[35] For instance, many people thought differently about the use of ethanol for fuel once information came out suggesting a link to increasing food prices. Another way to redefine an issue is to take advantage of unplanned external events that disrupt existing images. Focusing on events such as international crises or natural disasters may completely alter the image of an issue.[36] A recent example involves Hurricane Katrina, which forced policymakers to rethink the use of levees to structure waterways.

Advocacy coalitions contain journalists and other media actors. If such actors are at least somewhat sympathetic to a coalition's goals, they can be quite effective in using all these techniques to shape an image in ways that help engage and elicit the support of the larger public audience. As Baumgartner and Jones conclude, "issue definition, then, is the driving force in both stability and instability, primarily because issue definition has the potential for mobilizing the previously disinterested."[37] The degree to which proponents have been able to shape images favorable to repair efforts varies in the four cases examined here.

Economic Arguments. A second condition affecting the ability of change proponents to engage the larger public involves the economics of the proposed policy change. Advocates for a change are more likely to be effective if they can cite economic arguments showing that the benefits of their alternative policy exceed the costs of attaining it or the costs of maintaining the status quo. A paramount challenge for those seeking to remove the Glen Canyon Dam to restore natural flows in the Grand Canyon, for example, is to make the economic case that dam removal will not be excessively costly in terms of lost power or lost capacity to store water.

A variety of scholarly research supports the importance of economic arguments to policy change. The most obvious argument involves the use of benefit-cost analysis, which has had at least some impact on policy changes on issues ranging from dam construction to pollution control to economic regulation.[38] Other research involves case studies, particularly the literature on efforts to overcome collective action problems. Work in this field shows that a policy change to affect some collective goods problem is more likely to occur if advocates show that current patterns of resource use are costly or ineffective.[39] Another line of research comparing the behavior of the fifty states has often shown a statistically significant role for perceptions of economic conditions as well as actual economic conditions on state innovations and programs.[40]

Whether independently or in cooperation with pro-change advocates, the academics and researchers involved in advocacy coalitions also may make compelling arguments for policy change. If they do so, a pro-change coalition is more capable of eliciting support from members of the larger public who may otherwise be reluctant to back proposed changes with financial resources. A classic example of advocates using academics to make economic arguments to powerful effect occurred in the case of the Grand Canyon, although this instance involved a dam controversy prior to the current one involving Glen Canyon Dam. In the early 1960s, David Brower and others opposing dams in the Grand Canyon area recruited economists and engineers to argue that not only would the proposed dams be costly but that cheaper alternatives were available.[41] The impact of economic arguments in mobilizing public support can be especially important during times of economic stress, such as today, when many citizens are reluctant to consider new ideas that may entail new costs. Economic arguments also are crucial in cases involving environmental restoration when opponents of change argue that such projects are frivolous. The economic arguments in the four cases studied in this volume have not been uniformly compelling.

Scientific Evidence. A third condition affecting the likelihood of audience involvement in change efforts concerns the scientific evidence for the alternatives to the status quo. Evidence that a change to the status quo can be beneficial to society can be useful in enlisting public support for the change. For example, those seeking to reintroduce wolves to Yellowstone had to present biological arguments that the impacts on other species in the larger ecosystem would be positive.

The impact of scientific evidence on policy change proposals is not always clear, and the role of science in recent political debates on many issues has been controversial. Different sides often present their own versions and interpretations of scientific evidence. Nevertheless, a positive impact of scientific evidence has been apparent in numerous environmental policies, particularly with efforts to achieve collaborative changes to the status quo.[42] Again, consensus does not always exist on the science presented to support or reject a change. It is fair to say, however, that if those seeking change are unable to offer substantial scientific evidence to support their proposal, the lack of support only enhances the status quo. One classic example concerns climate change policies in the United States in the 1980s and 1990s. While the science on climate change has become increasingly confident and nearly unanimous in recent years, the lack of scientific consensus in earlier decades slowed U.S. action or at least provided a justification for inaction in spite of calls for new policies. Indeed, studies have shown that efforts to change policies that are not supported by "hard science" are not likely to succeed.[43] The four cases studied here vary in terms of the confidence inspired by available scientific evidence.

Agency Commitment. A fourth condition that makes audience involvement in and support for policy change more likely is if those who would manage the proposed changes—and especially those who would implement the changes—are committed and cooperative. If the larger public senses a lack of commitment from key agencies, it will be reluctant to become involved or to support change efforts. At the Everglades, for example, restoration efforts involve officials in not just an agency perceived as sympathetic to such goals (the NPS) but also the U.S. Army Corps of Engineers, an agency whose history involves more construction than restoration.

The literature suggests that agencies play a key role in policy formulation and implementation, although all public agencies are subject to potentially significant pressure and constraints from external institutional actors, such as Congress and the courts.[44] Empirical studies show that while agencies vary in terms of autonomy and responsiveness to other institutional actors, they do have some discretionary authority;[45] a theoretical framework of policy change therefore must take into account the motivations of those within key agencies. All agencies have some sense of identity or organizational culture or, as one scholar of bureaucracy described it, "a persistent, patterned way of thinking about the central tasks of and human relationships within an organization."[46] Inevitably, that sense of identity is linked to the agency's mission and the conditions surrounding its creation.[47] If the key agency involved in implementing a policy has an organizational culture that is antipathetic to change, then agency officials can use their discretionary authority to make significant changes to that policy less likely.

Having the commitment of the agency with primary jurisdiction over a policy is not enough, however. Managing proposed policy changes may well require the action of multiple agencies. That possibility increases when policies involve multiple jurisdictions, such as states, and multiple tasks, as do many restoration efforts involving large ecosystems. The importance of scale has been noted in policy efforts ranging from international treaties to collective action efforts involving common resources.[48] Having a mandate to perform multiple tasks—for instance, to create jobs and implement fiscal austerity measures simultaneously—inevitably complicates any policy.[49] In cases involving multiple jurisdictions and multiple tasks, as with the Everglades, multiple agencies often are involved in both formulating and implementing policy changes. If multiple agencies that have different organizational cultures are involved, then the potential for tension, if not conflict, is high. The policy literature, including classic implementation studies and recent work on multiple principals, describes the potentially problematic impact of the involvement of more than one agency in putting policies into place.[50] Communication and cooperation between the agencies are essential to overcoming such problems.

Therefore, another condition affecting the likelihood of public support for policy change involves the actions of agency officials. Agency officials who are committed to the goals of advocacy coalitions can make efforts to change policy more effective by coordinating the supervision of multiple tasks and reducing interagency conflict to make such changes more manageable. The attainment of policy goals is "unlikely unless officials in the implementing agencies are strongly committed to the achievement of these objectives."[51] Interagency commitment and cooperation vary in these cases.

Summary

To summarize, those seeking significant change to the status quo at the four parks analyzed here are more likely to attract needed support from the larger public if certain conditions are met. Positive issue definition, compelling economic arguments, convincing scientific evidence, and agency commitment are required if proposed changes are to occur. The working hypothesis for the following empirical examinations therefore is the following: When the coalitions seeking change at these parks effectively create and utilize those conditions, then they are more likely to achieve substantial change, in this context repair of damage from past behavior. The less effective pro-change forces are at creating effective coalitions and favorable conditions for engaging the larger public, the more likely efforts to repair damage will remain incremental and inconsistent.

This working hypothesis is logical and straightforward. It may not seem counterintuitive, but I will revisit the argument in the final chapter to present a more nuanced view based on the empirical assessment of the cases. As is, the general theoretical framework described above does facilitate the use of important policy change concepts in an accessible way; it also enables the explanation of outcomes. As other scholars have observed, much policy analysis focuses on process when what is needed is to understand outcomes: "We also need more analyses of distributions of outcomes though this has not been a favored mode of analysis in political science or public policy."[52]

Organization of the Volume

The next four chapters provide considerable background for each case, consistent with the need, stressed by the predominating theories of policy change, to analyze policies over decades of time.[53] However, I focus mainly on the most recent major decisions regarding repair efforts and the implementation of those decisions. That serves two purposes. Substantively, the major recent decisions determine what is happening in the parks today. Analysis of those policies allows us to be current and to anticipate future action (or inaction). Methodologically, the focus on recent decisions makes the cases more comparable in terms of their individual time frame. I argue that the outcomes in these cases can be explained by the actions of pro-change coalitions and the conditions that they create and use to achieve public involvement. Each case study therefore describes the behavior of those seeking changes and those conditions.

The case studies use whatever information is available. I have followed and in some cases participated in these repair efforts for years. I have studied and been involved with the parks for decades. Therefore, I use some personal anecdotes as well as historical reviews, archival records, and interviews with a wide range of people involved in each case. The participants in these interviews, at least the ones willing to be identified, are listed in the appendix. My personal anecdotes are not meant to be self-indulgent but rather to convey some sense of the character of these places. Indeed, understanding them is truly possible only by experiencing them on the ground (or water, as the case may be). I also attempt to be as analytical as possible, but nearly anyone who has spent much time in places such as those examined here develops certain views and feelings, and my own preferences become apparent.

The final chapter summarizes, synthesizes, and generalizes beyond the case studies. I argue that the lessons from these cases can be applied to other attempts at restoration and efforts to change policies. Obviously, these repair projects are not the only ones occurring in the national parks, nor are they even the only ones happening in these specific parks. These projects are, however, crucial to the future of these places, and they offer lessons to others who are attempting to rethink past public policies.

These bison have temporarily stopped traffic in Yellowstone's Lamar Valley. Yellowstone provides greater opportunities for wildlife viewing than almost any other natural setting in the world.

2

REINTRODUCING WOLVES AT YELLOWSTONE

Nowhere else in any civilized nation is there to be found such a tract of veritable wonderland . . . the wild creatures of the Park are scrupulously preserved; the only change being that these same wild creatures have been so carefully protected as to show a literally astounding tameness.

—PRESIDENT THEODORE ROOSEVELT, APRIL 24, 1902

The wolf recovery has been incredibly successful, but very difficult. The real challenge has been to reverse past policy.

—DOUG SMITH, YELLOWSTONE NATIONAL PARK WOLF PROJECT LEADER, 1999

Like nearly everyone else at the campground, I was camping at Pebble Creek for a specific reason. The reason was not the scenery, although Pebble Creek is a pretty spot, with its namesake brook bubbling down the hillside behind the campground to feed into the Lamar River in the valley below. Nor was the reason the amenities. The campground is rustic, complete with pit toilets, and showers are available only when it rains. Nor did I come to enjoy a long, peaceful night's sleep, since nearly everyone in camp stays up well past sunset and gets up well before sunrise. They do so for the same reason that I was camped there: we all wanted to see wolves.

Yellowstone has always been a magnet for those hoping to view wildlife. Both the Roosevelt and Smith quotes at the opening of the chapter carry the promise of seeing "wild creatures" in the park. Nevertheless, a century lies between those quotes, which, taken together, suggest that things at Yellowstone changed during that time. Perhaps the most important of the changes involved wolves. To recapture the remarkable state of nature that Roosevelt promised required reversing the public policy to which Smith referred. How did it happen?

Undesirable Inhabitants?

The Roosevelt quote aside, the desire to see wolves was not exactly common in the years leading up to the establishment of Yellowstone National Park in 1872 nor in the century that followed. Historically—and certainly well beyond the Yellowstone area—most people despised wolves. The villain in *Little Red Riding Hood* was only the most obvious manifestation of the fears of children, families, and livestock owners alike; many stories in medieval times even compared wolves to the devil.[1] Those fears, widespread in Europe and Asia, were part of the baggage that came along with explorers and settlers to the North American continent. Indeed, one can only imagine what John Colter and other early explorers of the Yellowstone area felt when, huddled around their campfires on dark and lonely nights, they heard wolves howling in the distance. Whereas most people today will hit the record button on a tape player to capture the sound, Colter and other early travelers were more likely to cock their Winchesters or at least throw another log on the fire.

"The Wanton Destruction of Fish and Game"

The issue of wolves at Yellowstone has always inspired debate, including arguments over whether they were even there before the park was established. In his comprehensive 1977 history of the park, Aubrey Haines summarized the feelings of many early explorers by praising Yellowstone as

> a great outdoor zoo preserving a more representative sample of the primeval fauna of the American West than is now found anywhere else. . . . Here, living under conditions very nearly those existing when white men first entered the area, are elk, buffalo, mule deer, moose, antelope, big-horn sheep, black and grizzly bear, cougar, coyotes, wolves, beaver and a number of smaller animals.[2]

Yet others questioned the inclusion of wolves in that statement. Park historians Paul Schullery and Lee Whittlesey therefore analyzed more than 168 accounts of the area written prior to 1882. They cited accounts specifically of wolves, including, for example, an 1870 report of a campsite in the Lamar Valley being "attacked by wolves."[3] Ultimately, they found an "overwhelming body of contemporary opinion in favor of wildlife abundance."[4] Perhaps the most convincing piece of evidence for abundant game in Yellowstone—and the corresponding presence of wolves—was the number of the latter killed in the park in the early part of the twentieth century.

In fact, wolves were often the targets of American hunters. Already hunted to extinction on many continents, wolves had made a home in North America, numbering well over a million before the Europeans arrived.[5] Their ability to colonize large parts of North America was consistent with their history; besides humans, wolves have been the most widespread land mammal in the world for centuries.[6] Most settlers in the western United States despised wolves and condoned if not participated in hunting, poisoning, trapping, and even torturing the animals. Their actions were driven by their fears, some founded on the realities of lost livestock and others, much less warranted, on tales of stolen children and waylaid travelers. This perception was pervasive in spite of the fact that, according to noted wolf expert and biologist David Mech in 1991, "There is no documented case of an unprovoked nonrabid wolf killing or seriously injuring a human being in North America."[7] These fears grew despite the more attractive attributes of wolves: these are animals that mate for life, care for their young, and are incredibly loyal to their pack. Nevertheless, government agencies, notably the U.S. Biological Service (forerunner of the Fish and Wildlife Service) pursued wolves vigorously. Between 1870 and 1877, for instance, government-sponsored hunters killed more than 1,800 wolves in thirty-nine national forests.[8]

The establishment of Yellowstone in 1872 did not at first stop the hunting in the park. Early survey parties and visitors to the park often were guided by hunters who had discovered the game in the area. Hide-hunters killed vast numbers of animals in the first decade of the park's existence.[9] In 1883, Secretary of the Interior H. M. Teller instructed the superintendent to stop sport and subsistence hunting within the park. While his order provided the authority to stop the hunting, when the U.S. Cavalry took over the police function in Yellowstone in 1886, it provided the means. As a result, the populations of some species grew immensely, and at least officially, park personnel did not kill any wolves between then and 1914.[10]

After the establishment of the National Park Service in 1916, however, agency personnel behaved quite differently. The Organic Act, which created the NPS, called upon the agency to "provide against the wanton destruction of the fish and game found within said Park."[11] However, the act also stated that the secretary of the interior "may also provide in his discretion for the destruction of such animals, and such plant life as may be detrimental to the use of any said parks, monuments, or reservations."[12] Notably, the 1883 list of protected species omitted carnivorous predators such as wolves and grizzly bears.[13] Arguably, NPS managers saw wolf hunting as a means to build good relations with local livestock owners and to increase the new agency's perceived value and its budget.[14] The NPS therefore condoned and even sponsored the killing of wolves in Yellowstone until no viable populations were left in Yellowstone by the mid-1920s.[15] As one critic concluded in 1998, "the Park's reputation as a great game sanctuary is perhaps the best-sustained myth in American conservation history."[16] For the most part, by the 1930s gray wolves were abundant only in remote areas of Canada and Alaska. The dates of park creation and wolf extinction, along with others related to wolf management policy, are shown in table 2-1.

Where Necessary Recreated

It is not surprising, in retrospect, that the absence of wolves caused significant changes to the Yellowstone ecosystem. Without natural predators, the elk population grew substantially, in spite of the fact that the NPS killed or removed more than 58,000 elk between 1935 and 1961.[17] That growth had severe consequences for stream banks, vegetation, and wildlife, such as beavers, that were dependent on the vegetation. Scientists began to notice. A group led by George Wright published a study of wildlife resources in 1933 that linked predator behavior to the size of ungulate populations and concluded that protecting predators could benefit the larger ecosystem.[18] By the 1940s, ecologists such as Aldo Leopold were more emphatic, stating that predators were an important part of the natural processes within ecosystems. In *A Sand*

Table 2-1. *Timeline of Actions Related to Policy Changes at Yellowstone*

Year	*Policy action*
1872	Establishment of Yellowstone National Park
1916	Establishment of the National Park Service
1925	Viable wolf populations reported to have been eliminated
1971	First interagency meeting for management of wolves held in Yellowstone
1973	Passage of Endangered Species Act; wolves listed
1980	FWS completes Northern Rocky Mountain Wolf Recovery Plan
1987	FWS completes revised Northern Rocky Mountain Wolf Recovery Plan
1987	Bill to reintroduce wolves to Yellowstone fails in Congress
1988	Congress directs NPS and FWS to conduct studies for reintroduction
1989	Bill to reintroduce wolves to Yellowstone fails in Congress
1990	Bill to reintroduce wolves to Yellowstone fails in Congress
1991	Congress directs FWS and NPS to prepare draft EIS on wolf reintroduction
1992	Congress mandates completed EIS by January 1994
1992	FWS and NPS conduct hearings and scoping process for EIS
1993	Draft EIS completed and presented for comment
1994	Final EIS released
1995	Wolves released into park
2000	Circuit court ruling allowing wolves to stay in park
2008	Bush administration attempt to delist wolves

Source: U.S. Fish and Wildlife Service, *Final Environmental Impact Statement for the Reintroduction of Gray Wolves to Yellowstone National Park and Central Idaho* (Washington: 1994), appendix 1.

County Almanac, Leopold wrote that seeing a "fierce green fire" dying in the eyes of a wolf after he shot her had contributed to his conversion to being an ecologist.[19] In 1944, Leopold recommended reintroducing wolves to Yellowstone.[20] In a report to Congress in 1963, Leopold's son A. Starker Leopold and others recommended that "biotic associations within each park be maintained, or where necessary recreated, as nearly as possible in the condition that prevailed when the area was first visited by the white man. A national park should represent a vignette of primitive America."[21] The Leopold report became a seminal document for park management.[22] In 1966, several other biologists recommended restoring wolves specifically to Yellowstone.[23] One of the most prominent recommendations came from biologist and wildlife expert John Weaver, who called explicitly for a wolf reintroduction program in a report to the NPS in 1978.[24]

Scientific arguments about possible benefits from a viable wolf population had some resonance within the National Park Service and the U.S. Fish and Wildlife Service. At an interagency meeting in Yellowstone in the fall of 1971, officials of different federal and state agencies discussed possible actions regarding wolves, including reintroduction. Reports at the time suggested that as many as ten to fifteen wolves already had wandered into the Yellowstone area.[25] Two years later, the Endangered Species Act (ESA) called for the protection and recovery of endangered species "solely on the basis of the best scientific and commercial data available" regarding biological, not economic, issues.[26] Further, the ESA mandated a recovery plan for all listed species. Therefore, after the Northern Rocky Mountain wolf was listed as endangered in 1973, the FWS called for a recovery plan in the region along with compensation to local farmers and ranchers for livestock losses. The NPS concurred with the FWS suggestion. In the 1974 master plan for Yellowstone, park officials recommended natural regulation of ungulates: "An important element in this approach is the reestablishment of natural predators within the range of the northern Yellowstone elk herd."[27] Thus were planted the early seeds of an alliance between biologists and agency officials.

The possibility of introducing wolves in the greater Yellowstone area inspired systematic analyses in the 1980s. The FWS completed a recovery plan for the Northern Rocky Mountain wolf in 1980 and then issued a revised plan, following comments, in 1987. Occasional anecdotal reports of sightings suggested that wolves still found their way into the park or that some NPS personnel had secretly reintroduced them.[28] Other observers suggested that wolves would eventually make their way to the park on their own, wandering down from Canada or northern

Montana.[29] By the early 1980s, wolves had begun reestablishing a presence in northern Montana, particularly around Glacier National Park. The general conclusion was, however, that even if solo wolves wandered or were ushered into the area, at the onset of the 1990s "no viable wolf population remains within the Yellowstone ecosystem."[30]

Responding to Livestock and Agricultural Interests

Even with the increased interest among some wildlife officials in establishing a viable population of wolves, attaining formal political support for reintroduction would not be an easy task. Livestock owners in the area objected to the idea intensely, and their objections registered with and were echoed by their representatives to Congress. As Yellowstone superintendent Mike Finley once told me, "Most of the time, all members of Congress do is respond to constituents and around here that means cattle and agriculture."[31]

Nevertheless, the wolves also had their champions in Congress. Representative Wayne Owens (D-Utah), a moderate from Salt Lake City with a strong environmental record, took up the wolf cause in the late 1980s. Owens's views represented those of the 2nd congressional district, the "least Mormon and most cosmopolitan part of Utah."[32] His first effort, in 1987—a bill requiring the NPS to bring wolves back to Yellowstone—did not reach the floor. In 1989 he proposed in another bill that an environmental impact assessment be done of a plan to reintroduce wolves in order "to restore ecosystem balance." That bill did not make it out of committee (hearings were held), largely because it lacked the support of the Wyoming delegation.[33] Then, in 1990, Senator James McClure (R-Idaho), a conservative known for taking unpredictable positions, introduced a bill to reestablish wolves in Yellowstone and central Idaho, but this bill also went nowhere.[34] Some suspected that McClure's motivation was the fact that wolves had already been sighted in Idaho, where, as a naturally occurring population, they would receive the full protection of the Endangered Species Act.[35] His bill called for reintroducing gray wolves in both Yellowstone and parts of Idaho but also removing them from the endangered species list in other areas of the country, including, presumably, parts of Idaho.[36]

That those fledgling efforts failed is not surprising. Arguments against wolf reintroduction in Congress were emotional and intense. For example, in a 1989 House of Representatives hearing, Representative Don Young (R-Alaska) vilified the wolf and then warned, "God help the poor cattle rancher and the sheep rancher when he tries to protect his stock." Senator Conrad Burns (R-Montana) made the same argument, emphatically stating "stockgrowers and farmers are very concerned about this proposal to reintroduce."[37] Others warned of potential harm to humans. Senator Steve Symms (R-Idaho) said that wolves "pose a real danger to humans." And Senator Burns was more graphic, warning that if wolves came back to Yellowstone, "There'll be a dead child within a year."[38]

One of the major arguments about the wolf issue involved who would actually do the reintroduction. When they have banded together, senators from Western states have formed a powerful voting bloc that has stymied legislative efforts to change policies—for example, by increasing grazing fees or revising the 1872 mining law. Many of their arguments use and build on the resentment toward federal authorities that has been a tradition in the West since the late nineteenth century. Wolf proposals, then, involved not just the renewed presence of a predator but reintroduction by the federal government. As historian Gary Ferguson writes, "while livestock may be the kindling, the real fuel for the burn is that the federal government is behind the project. . . . Westerners couldn't buy a more perfect evil if they rode straight to hell with a saddlebag full of cash."[39] Or, as one stockman said in 1989, "It's not so much wolves we're afraid of, it's wolf managers."[40]

Another aspect of the argument against federal intrusion involved the boundaries of the park. When first established, the boundaries had little to do with wildlife protection. Authorities had drawn the original boundaries from rough notes and maps from the 1871 Hayden survey of the area. They designed them to include the canyons, waterfalls, hot springs, and geysers that would be important in attracting tourists to Yellowstone, which was the goal of the Northern Pacific Railroad and others. Even later changes to the boundaries had less to do with the migratory or territorial needs of wildlife than with not encroaching on neighboring private interests or national forest lands. Conservationists, notably John and Frank Craighead, had long pushed for new borders to include a more natural ecosystem. In 1983, partly in response to fears that the Reagan administration might open up neighboring lands for development, political science professor Ralph Maughan and several other activists formed the Greater Yellowstone Coalition to advocate specifically for a Greater Yellowstone Ecosystem (GYE).[41] Such an ecosystem would vastly expand the 2.2 million acres of the park. Therefore, in the wolf debates, many opponents characterized the reintroduction plan as a diabolical plot to expand the park's boundaries. Representative Young testified in 1989 that "my firm belief is this [reintroduction] is an intent to extend the borders of Yellowstone Park."[42] Similarly, Senator Alan Simpson (R-Wyoming) argued that "if you were really just staying with Yellowstone, it might not be quite so bad, but you're not. You are using the concept of the Greater Yellowstone Ecosystem, which has not been accepted well in the West because it goes clear out through various forests, and BLM, and the whole works. It just goes on and on and on."[43] Senators from Wyoming and Montana thus fought wolf reintroduction by placing holds on bills that would have authorized federal studies.

The opposition to wolf restoration proposals remained powerful through the preliminary discussions of the 1980s and into the 1990s. In addition to the traditional support of livestock ranchers, strong support for opponents of reintroduction came from organizations such as the Wise Use Movement, a loose-knit coalition of groups advocating resource extraction on public lands and property rights.[44] Furthermore, many officials in neighboring states opposed reintroduction and also were suspicious of any discussion of an expanded ecosystem, and they too contributed to efforts to reject such proposals. As Bob Ekey of the Greater Yellowstone Coalition told me in 1996, "the states exert an incredible amount of influence over the park."[45] Not surprisingly, then, in 1988 and 1989 Congress refused to authorize the environmental impact assessment regarding wolf reintroduction.[46]

Replacing Myth and Folklore

In spite of the opposition described above, the reintroduction idea gained some momentum during the 1980s. The seeds had already been sown, as discussed earlier, with the arguments of scientists and some agency officials for the benefits of restoring a more natural balance. That momentum produced an advocacy coalition that included scientists, agency officials, and interest groups—the kind of coalition described in the first chapter as potentially quite persuasive.

Through the 1980s, scientists grew increasingly explicit about the potential value of reintroducing wolves to the ecosystem. The increasing size of the elk population continued to raise the question of imbalance. Further, wolf studies provided compelling data that wolves did at one time inhabit Yellowstone, do not eliminate their prey populations, and would not devastate livestock around the park if reintroduced.[47] By 1991, David Mech could conclude that the "scientific data about wolves and wolf predation began to replace myth, legend, folklore, and frontier thinking."[48]

Pivotal public agencies warmed to the goals of reintroduction, albeit at different rates and with different degrees of explicitness. In 1987 the Fish and Wildlife Service proposed a recovery plan for the Northern Rocky Mountain wolf that called for release of wolves in large and remote areas as well as efforts to reduce potential conflicts with humans and livestock.[49] The Forest Service, responsible for many of the lands neighboring Yellowstone, was less supportive of the idea. The agency's emphasis on multiple uses of public lands, including grazing in national forests, made their personnel more sympathetic to the demands of ranchers and wary of reintroducing wolves. The NPS was somewhat inconsistent. William Penn Mott, director of the agency from 1985 to 1989, was a staunch proponent of wolf reintroduction in spite of the conflicting views of many within the Reagan administration. Mott's successor was much more guarded, however. In 1989, NPS director James Ridenour testified that the NPS did not support efforts to expedite the process but rather promised "due consideration of the impacts on both human populations, domestic livestock and wildlife populations, and not without consultation with and substantive input from affected state wildlife management agencies, state elected officials, and members of Congress."[50] Nevertheless, the 1990 NPS plan for the Rocky Mountain region stated that "reintroduction of the species will help provide an ecological balance to the area."[51]

Environmental interest groups were a significant source of pressure for reintroduction. Some conservation groups, notably the Wolf Fund, founded by Renee Askins in 1986, were formed specifically to push the proposal. Many national groups were quite active in their support. Most notably, the Defenders of Wildlife began raising money for a trust fund that could be used to compensate ranchers for livestock losses.[52] The Greater Yellowstone Coalition, which since its beginning in 1983 had grown to a membership of more than 7,000 individual members and 120 member organizations by the mid-1990s, consistently pushed the concept of the GYE and also provided a strong voice for wolf reintroduction.[53] The pressure for reintroduction was growing.

The Coming of the Wolves

As I settled into the Pebble Creek campground, I thought about how I had become like so many other visitors to the park in my determination to see the wolves. I had been coming to Yellowstone for decades, but like most other visitors in the 1970s and 1980s, I didn't think much about the lack of wolves then. Since 1995, however, I had tried several times to see the animals, on several different visits to the park, and finally realized that I wasn't going to just stumble into a random encounter. I needed to increase my chances by camping out in the Lamar Valley. Furthermore, I had tried to time the trip to maximize the possibility by learning about the wolves' hunting habits. In late May, the elk had not yet moved to the high country. They still were enjoying the vegetation at the lower elevations—and thereby providing a tempting target for hungry wolves. Snow still covered the surrounding peaks, a storm having moved through just the day before I arrived at the park.

Not only was I willing to stay as long as necessary, I was ready to take chances if I had to. Therefore I was not surprised when, sitting at a picnic table eating a peanut butter sandwich, I got a sober reminder that the Lamar Valley is not a risk-free environment. A hiker coming down the trail from the hill behind the campground paused briefly at my picnic table to catch his breath and issue a brief warning: "Just so you know," he said matter-of-factly, "there's a grizzly bear coming down this trail about a minute behind me." "Ah well," I figured, as he hustled off and I hurriedly swallowed the last of my lunch, "the grizzlies come with the territory, even now with more wolves back in the ecosystem."

To Establish an Experimental Population

Congress laid the groundwork for reintroduction even before the pro-wolf Clinton administration took office in early 1993. In 1991, the House Appropriations Committee approved funds for an environmental impact assessment of reintroduction. Facing opposition in the Senate, House conferees, using compromise language crafted by Representative Owens and others, agreed to spend money on research without committing to actual reintroduction at this point.[54] In the Senate deliberations, Senator Simpson acknowledged that reintroduction was likely but stressed the need to do it in a way that his constituents would accept: "If Congress is going to forge ahead with wolf reintroduction . . . I believe we can honestly deal with this issue and find a solution that will be acceptable to all sides."[55] The conferees appropriated $200,000 for wolf research and called for an assessment to be produced within eighteen months. Amendment 218 of the Department of the Interior (DOI) appropriations bill directed the secretary of the interior to "appoint a 10-member Wolf Management Committee. The Committee's task shall be to develop a wolf reintroduction and management plan for Yellowstone National Park."[56] The National Park Service and the Fish and Wildlife Service then appointed a committee to develop a plan for reintroduction.[57]

Thus, when the Clinton administration took office, the studies for reintroduction already were under way. The new administration, especially Secretary of the Interior Bruce Babbitt, encouraged and supported the studies.[58] In the 1994 DOI appropriations bill, Congress authorized $40,000 "for the research program relating to habitat and repopulation studies and possible interactions between wolves and mountain lions in and around Yellowstone National Park."[59] The bill contained many provisions and thus defies precise characterization of the split of votes on wolf reintroduction. Nevertheless, to give some idea of the controversial nature of the overall appropriation, the vote in the House was 278 to 138, split largely on partisan lines (Democrats, 227–17; Republicans, 50-121).

The agencies now had the funds needed to begin to take action. The FWS embarked on an extensive process that began with a notice of intent to prepare an environmental impact statement (EIS) in April 1992. Personnel gathered information, reviewed existing evidence, solicited input, and held open houses; nearly 4,000 comments were received during the scoping process alone, which was designed to develop plan parameters.[60] The draft EIS was completed in June 1993 and the final EIS in May 1994. The entire process involved thousands of letters sent to potentially interested participants, scoping brochures outlining possible actions sent to tens of thousands of people, a planning brochure inserted in thousands of Sunday newspapers in the three neighboring states, more than 150 public hearings, and analysis of more than 160,000 public comments, more than had ever been received for any such action.[61]

The EIS discussed five alternatives and identified one as preferred. The nonpreferred options were natural recovery (allowing the wolves to return on their own), prevention of wolf recovery, leaving the issue to the states, and reintroduction of wolves without assigning them experimental status. The preferred alternative was to reintroduce wolves as a nonessential experimental population. The Endangered Species Act states that the secretary of the interior can authorize experimental populations but "only when, and at such times as, the population is wholly separate geographically from nonexperimental populations of the same species."[62] Experimental status means that wolves do not receive the full protection of endangered species but can be shot outside the park if they threaten livestock. The EIS proposed to "establish an experimental population" and promised "management of wolves by government agencies and the public to minimize conflicts on public lands, effects on livestock, and impacts on ungulate (deer, elk, etc.) populations."[63] In other words, the wolves would arrive with a special status: they would not be listed as fully endangered and their legal protections were limited, but nevertheless they would be in the park.

The agencies defined the recovery goal as the successful breeding for three successive years of a minimum of ten pairs of wolves in each of the three recovery areas (the Greater Yellowstone Ecosystem, Idaho, and Montana). They predicted a recovered wolf population in Yellowstone of more than 100 animals by 2002. Program officials envisioned the ultimate interbreeding of the reintroduced subpopulations with the populations recovering in northern Montana, which would form a "metapopulation" in the northern Rockies and allow for the eventual removal of the gray wolf from the endangered species list.[64]

The experimental status proposal was a compromise between according full protection to the wolves and having no wolves at all. Not surprisingly, it inspired protests from livestock interests for being too protective and from some environmentalists for not being protective enough.[65] The wolves themselves, somewhat ironically, helped persuade both sides. By late 1992, various sightings within the Yellowstone area suggested that the wolves were returning on their own. A filmmaker recorded what looked like a wolf feeding on an elk carcass in the Hayden Valley. Visitors reported seeing paw prints. In September, a hunter shot a wolf just south of the park in the Teton wilderness area, although no one knew where it came from.[66] In a front-page story in the *New York Times,* FWS personnel estimated that about fifty wolves were already roaming the state of Montana.[67]

Many environmental groups recognized that experimental status was nevertheless a necessity, admitting that wolves would have a hard time making it to the park on their own in large enough numbers to create a viable population.[68] Even those who had argued for letting wolves return on their own and thereby gain full protection under the ESA admitted by 1993 that "most wolf advocates have rallied around the strategy of reintroducing wolves into Yellowstone under an experimental, nonessential designation."[69] The livestock community also began to think that the experimental compromise might be better than having naturally arriving animals receive full ESA protection. Not only would the wolves not receive full protection but, as the FWS observed in its environmental impact statement, "having wolves designated as a nonessential, experimental population under the FWS proposal means no critical habitat could be designated for wolves and none would be needed."[70] As NPS chief of resources John Varley recalled several years later, "senators from Idaho, Montana, and Wyoming were telling their ranching constituents that the wolves were coming and they needed to find middle ground."[71]

The prospect of wolves in Yellowstone represented a potentially remarkable change of policy. Those seeking the action had created conditions that were extremely effective at engaging the larger American public to support the proposal:

Framing the Wolf as a Symbol

The possible reintroduction of wolves to Yellowstone made for a vivid story. Here was a project that involved an animal that was both one of the most reviled and one of the most esteemed in world history, the flagship U.S. national park, and battle lines between the opponents as distinct as wolf tracks in the snow. Over the years leading up to reintroduction, the wolf's image changed from one of hated predator to noble ecosystem savior to a symbol of what had been lost in America. That change fostered significant public support for reintroduction.

The shifting of the image of the wolf from villain to hero took decades, but it accelerated from the 1960s to the early 1990s. Aldo Leopold's realization, described earlier, that a wolf was much more than just a moving target, resonated with many who saw the author as an ecological pioneer. The shift also grew out of comments from scientists that described wolves as "strong, intelligent, keen, and dynamic."[72] But the shift was based on more than science. Wolves captured the imagination of authors such as Farley Mowat, Roger Peters, and Barry Lopez, who dedicated his classic *Of Wolves and Men* to the animals.[73] Other authors described the wolves in spiritual terms; one poet, for example, recalled a wolf's howl as "the oldest of legends, the saddest of songs."[74] Hollywood got in on the act, reviving Mowat's *Never Cry Wolf* from 1963 to make a movie twenty years later by the same title. Kevin Costner's *Dances with Wolves* won seven Academy Awards in 1990, including the award for best picture. As one journalist noted, "Once the symbol of lechery and deception, the wolf has recently gained a wide following. . . . Some hunters and ranchers blame Hollywood for the wolf's current good standing. Movies like *Never Cry Wolf* and *Dances with Wolves* have made *Canis lupus* out to be much more cuddly than Little Red Riding Hood's stalker."[75] Or, as one rancher said at the time of reintroduction, "Now, only one and a half percent of the population is involved with agriculture and the other 98-plus percent have been indoctrinated by a Disney mentality about the wolf."[76]

As accurate as opponents of reintroduction might find that characterization, the wolf was framed as more than just a cuddly denizen of the wilderness. The wolf became a symbol of much larger issues, in particular the wilderness—physical or psychological—that remained in the United States. Between the summer of 1989 and fall of 1991, for instance, three major U.S. news magazines (*Newsweek, Time,* and *U.S. News and World Report*) published articles on the possible reintroduction, complete with eye-grabbing photos of wolves in natural settings.[77] *Newsweek,* in a cover article entitled "The Return of the Wolf," suggested that wolves might be an "inspiring symbol of the wild."[78] The *U.S. News and World Report* article concluded that with restoration, "a little more wild will be back in America's premier wilderness."[79] Reintroduction proponents used similar language. Hank Fischer of Defenders of Wildlife also saw the wolf as a symbol, a test to see "how willing humankind is to share this planet with other forms of life."[80] Advocate Renee Askins said that wolves would be reintroduced because "Westerners respect wild things, and the wolf is one of the wildest things;"[81] she later added that the reintroduction was not about saving wolves or about tourism but rather a "symbolic act" reflecting a new approach to wilderness.[82] Chief of resources Varley said in 1990 that "emotionally and philosophically, it's important that we have one place in the lower 48 with all the species that were on these shores when white men came to this country."[83] Or, as NPS wolf project leader Doug Smith concluded years later, "wolves are no longer just a biological issue, they're a social issue."[84]

Livestock and agriculture groups encouraged schools to discuss the wolves in a less sympathetic light than the one shining in the popular press.[85] Some ranchers and other opponents of reintroduction also saw the issue as symbolic, but of a different issue. "Ranchers say keeping the wolf out of the West is a symbolic stand for control. . . . With restoration, they fear, will come a host of regulation."[86] In a lawsuit following reintroduction, one of the lawyers for those seeking to end the program argued that the issue involved not just wolves but also fundamental questions of land use: "Once you get wolves in there and let them run loose, pretty soon the whole Rocky Mountain area will be locked up."[87] For opponents, the wolf was indeed a symbol, not of wilderness but of federal government control of the West.

The former image prevailed. As one journalist concluded prior to restoration, "now, even in their own communities, ranchers seem to be losing the spin battle."[88] The story was too powerful, the battle a symbol not just of reclaiming the wilderness but also of redemption. The 1991 *Newsweek* article, resplendent with stunning photographs, concluded by stating, "What man had destroyed, man may redeem."[89] One reporter said of the reintroduction that "the project is viewed as a grand restorative deed, balancing a zealous act of one era with a corrective act of another."[90] After the wolves were back in Yellowstone, a 1995 *Newsweek* article referred to the reintroduction as "the Return of the Native" and concluded that "the mythic gray wolf . . . had come home."[91]

The framing of the reintroduction as a dramatic revitalization of the premier U.S. wilderness ecosystem did not occur by accident. Those supporting the program, from the agencies to the environmental groups, made a concerted effort to educate and convince the larger public of its importance. As Doug Smith said, "the public outreach for this program in the early 1990s was arguably the most ever done for a wildlife project."[92] Smith acknowledges that at least some segments of the public, notably ranchers, perceived the framing as one-sided and the process as not participatory enough, with the agencies mandating the path chosen for reintroduction. But he also argues that more stakeholder involvement probably would not have made the anti-wolf people any happier with the final outcome.

Science instead of Old Wives' Tales

The scientific arguments for reintroduction were less philosophical than those for issue redefinition, but they were just as compelling. The larger context for the discussions involved the protection and restoration of endangered species. The Endangered Species Act mandates that decisions be made on scientific grounds alone. Not only is that provision clear, but the law is quite explicit in terms of processes and actions. As one review of U.S. environmental laws states, the ESA may well be "the most powerful natural resources law in the nation or, for that matter, in

Without wolves as predators, the elk population in Yellowstone increased dramatically. Today, after the reintroduction of wolves, the size of the elk population is smaller, but many of them can still be seen in developed areas such as Mammoth Hot Springs, where they may go hoping to get away from predators.

the world."[93] That does not mean that it is not controversial. Although the ESA was passed almost unanimously in 1973, implementation immediately inspired controversy—first, with the snail darter case—and the controversy grew with spotted owls and hundreds of other species.[94] But the requirement that science determine decisions has remained intact.

In the Yellowstone case, the scientific and academic communities rallied around the argument that reintroducing wolves could restore the natural balance to the ecosystem. One example of the imbalance that was at least theoretically attributable to the lack of wolves was the increase of the elk population. The elk population tripled in size between the 1960s and the 1990s.[95] Agency personnel attempted intensive management and planned hunting to keep some limits on the herds. They also relied on natural regulation. For instance, during the winter of 1988–89, roughly 5,000 elk died of starvation; in contrast, the massive fires of 1988 killed less than 500 of the animals.[96] One article summarized the situation vividly: "But the area around Yellowstone National Park is overcrowded with big game and could use a natural hunter like a wolf. Three years ago, one of the more pathetic sights amid the grandeur of Yellowstone was that of hundreds of starving, dazed elk stumbling around towns on the border of the park, scrounging for food."[97]

Even after the deaths due to starvation and fire, the elk herd remained too large for the vegetation in much of the park. Biologists argued that wolves would not only cull the herd but also prey mainly on the old and injured members of the population. By 1992, Yellowstone superintendent Bob Barbee was willing to argue with those fearing heavy livestock losses by claiming that wolves would be less tempted to wander outside the park now that the area held more deer and elk than at any time since white settlers first moved to the area.[98] Indeed, when the wolves were eventually released three years later, the GYE contained roughly 4,000 bison, 30,000 deer and 56,000 elk.[99] Wolves would not only reduce the elk population, thus freeing up food sources for other grazing animals, but elk carcasses from wolf kills could provide a food source for animals like bears.[100] Research from other wolf-inhabited areas showed that smaller predators, such as wolverine and lynx, also benefited from the carcasses of the elk and deer that the wolves killed.[101]

The wolf was the missing ingredient in the ecosystem. Advocates, journalists, and park officials referred to the wolf's return as making Yellowstone "whole" again.[102] Others described the wolf as being the "only native animal missing from Yellowstone,"[103] and *U.S. News and World Report* concurred: "But one native creature is missing: wolves."[104] Park officials also argued that "reintroducing the predator will restore natural balance to the park's wildlife."[105] One account asserted that "now, most Federal biologists say eliminating the wolf was a grievous error, upsetting the balance of nature in one of the few places in America that was balanced."[106]

Those arguments made sense to many scientists, including some of the most prominent wolf biologists in the world. As one account stated, "biologists see predators as balance wheels in ecosystems. No wolves mean too many elk, which is what Yellowstone has today."[107] Wildlife experts like Dave Mech and John Weaver endorsed the idea of reintroduction. Mech asserted that carnivores like wolves "do play important and unique roles in the natural functioning of ecosystems."[108] Weaver reiterated the restoration perspective, saying that "by bringing a native species back, we're making parks whole again."[109]

Reintroduction proponents could also point to evidence from other areas to buttress their arguments. Evidence from Minnesota indicated that wolves and livestock could live in the same area largely because wolves preferred wild animals such as deer to cattle or sheep.[110] The agencies had pursued similar reintroduction programs, such as those bringing red wolves to the Great Smoky Mountains and the Alligator River, North Carolina, area and, to a lesser extent, Mexican gray wolves to the Southwest. Those programs produced mixed results, but they did result in some benefits and provide some empirical lessons.[111] In summing up the overwhelming scientific evidence, Mike Hayden, a biologist and a former assistant secretary of the Department of the Interior, declared, "Science must prevail over all the old wives' tales."[112]

Scientists also weighed in on the goals of reintroduction. Would the stated goal of successful breeding for three consecutive years by ten pairs in three areas create a viable population? A survey of forty-three wolf biologists showed that two-thirds believed that the goal would meet at least minimum standards. They concluded that more wolves would be helpful for genetic diversity but that the plan was "reasonably sound and would maintain a viable wolf population in the foreseeable future."[113]

Finally, if for no other reason, many scientists supported the program for its research value. As one noted ecologist wrote, "ecological-process management [at Yellowstone] has provided an understanding of how the ecological systems function in natural parks, which will assist with wildlife management inside and outside national parks."[114] Many biologists and ecologists expressed excitement at the chance to see how bringing a predator back could impact an ecosystem.[115]

The Economics of Wolves

Positive issue definition and persuasive scientific arguments were not enough to guarantee strong public and political support. The coalition favoring wolf reintroduction also had to try to educate and convince the broader public and even some traditional opponents that the program was worth pursuing.[116] Economic arguments were compelling for several reasons.

Simply stated, the project promised more benefits than costs. Characteristics of the Greater Yellowstone area in the mid-1990s are shown in table 2-2. As the table shows, the anticipated livestock losses from wolf reintroduction were minimal compared to losses from other causes. The table presents a range, but expected annual losses due to predation by a population of 100 wolves amounted to nineteen cattle and sixty-eight sheep.[117] The anticipated benefits from increased visitation were much higher than expected losses to ranching and hunting interests. "Existence value" refers to the value of knowing that a resource exists; it is calculated by asking how much people would be willing to spend to know that a resource is there even if they never use it. For wolf recovery, that amount was estimated to be in the millions of dollars.[118] Finally, the FWS and NPS anticipated annual costs of $616,000 for managing the program and $316,000 for monitoring for the first five years.[119] In other words, the cost to taxpayers was less than $1 million a year.

Several further points made the economic analyses even more persuasive. First, even if the fuzzier parts of the projections (existence value) are not included, the anticipated benefits from the project far exceeded the anticipated losses. That conclusion was not just the view of zealous bureaucratic advocates. A University of Montana study estimated a total net social benefit to the GYE, mainly through increased tourism, of about $110 million over a twenty-year period. Other estimates predicted as much as $19 million a year in tourism benefits.[120] Supporting evidence involved changes in the region's economy. Regional economic activity had shifted away from an emphasis on extraction of resources to tourism and related amenities. Between 1969 and 1989, 96 percent of new jobs and 89 percent of the growth in labor income occurred in areas other than agriculture and extraction industries. Put another way, in 1969, one of every three employees in the region worked in resource use; by 1989 that figure was one in every six. As table 2-2 shows, the livestock business was now a small, albeit loud, minority in terms of contribution to the local economy.

Table 2-2. *Characteristics of the Greater Yellowstone Area, Mid-1990s*

Greater Yellowstone area size	64,800 square kilometers
Yellowstone National Park size	9,000 square kilometers (2.2 million acres)
Federal ownership of land	76 percent
Recreational visits per year to federal land	14.5 million
Regional population	288,000
Local total income	$4.2 billion
Local farm income	6.4 percent (55 percent livestock)
Peak numbers of cattle	354,000
Peak numbers of sheep	117,000
Total cattle mortality per year from all causes	8,340 (2.4 percent)
Total sheep mortality per year from all causes	12,993 (11.1 percent)
Anticipated cattle mortality from wolves per year	1–32
Anticipated sheep mortality from wolves per year	17–110
Anticipated ungulate mortality from wolves per year	1,200
Anticipated annual losses from reduced hunting	$404,000–$879,000
Anticipated annual livestock losses	$1,888–$30,470
Anticipated annual visitor increase	5–10 percent
Anticipated annual increased visitor spending	$23 million
Anticipated annual existence value	$8.3 million
Anticipated annual management costs	$615,500
Anticipated annual monitoring costs	$316, 250

Sources: Edward F. Bangs and Steven H. Fritts, "Reintroducing the Gray Wolf to Central Idaho and Yellowstone National Park," *Wildlife Society Bulletin* 24, no. 3 (1996); U.S. Fish and Wildlife Service, *Final Environmental Impact Statement for the Reintroduction of Gray Wolves to Yellowstone National Park and Central Idaho* (Washington: 1994), Abstract, chapter 4, appendix 5.

Second, while any livestock losses can be tough on ranchers, compensation funds did exist to mitigate the damage. By the time of reintroduction, Defenders of Wildlife had more than $100,000 in its fund to use as possible compensation for livestock losses. Ranchers were not swayed by that argument. As one said, "They're not obligated to pay us. Once the wolves are in there, they could run out of money. Then what?"[121] However, by 1995, Defenders had already compensated ranchers in Montana and Alberta with $17,000 for losses from naturally occurring wolves since 1987.[122]

Third, the proposed experimental status for the wolves provided a ready reply to any concerns about substantial losses in the livestock community. Analysts, opponents, and proponents of the project characterized the possibility of killing problem wolves as crucial. Citing the wolves' experimental status, one analyst said the reintroduction "incorporates the most people-pleasing compromises in the [Endangered Species] act's 22-year history."[123] Senator Simpson warned that full protection of wolves under the ESA would be problematic: "We have to have ways to take [kill] them if they leave the park."[124] The administration, notably Secretary of the Interior Babbitt, argued that the experimental plan restored wolves while protecting livestock interests.[125] Superintendent Finley noted that NPS authorities welcomed the management flexibility that came with the experimental designation.[126] Wolf project leader Smith stated that a major purpose of the experimental status for wolves was that "it relaxed some of the stiffness of endangered species protection."[127] Or, as NPS chief of resources Varley said, "We gave the livestock industry everything except no wolves."[128]

Although no one could know it at the time, the actual benefits of the program would exceed those anticipated in the EIS. Those figures are presented later in the chapter; suffice it to say here that even in the mid-1990s, economic arguments for reintroduction were compelling.

Agency Enthusiasm

In addition to the importance of a positive image, convincing science, and compelling economic arguments, the history of the wolf reintroduction, at least during the final planning and implementation stages, reveals the importance of the commitment and cooperation between the two lead agencies involved in the project. Interagency tension in the Yellowstone restoration was virtually nonexistent compared with that in the other projects described in this volume.

NPS and FWS personnel worked together closely throughout the planning, EIS preparation, and catch-and-release stages of the project. That was not that surprising. The interest among NPS personnel in repair projects such as wolf reintroduction, particularly in recent years, has been well documented.[129] NPS wolf project leader Doug Smith savored the prospect of making a real difference. The history of the FWS differs from that of the NPS, but it also is one that led to the likelihood of staff enthusiasm for the wolf project. Historically the FWS has received less attention and support than the NPS, even while being given challenging mandates such as that of protecting endangered species. The result has been an agency that is usually "overwhelmed by inadequate resources, bitter controversies, and difficult if not impossible assignments."[130] Like many people working in the NPS, however, many of those working in the FWS are dedicated to the mission of protecting species and thus, despite difficult conditions, "the Service has done a heroic job."[131] Thus, FWS personnel relished the chance to assist a species recovery program that was gaining popularity and support.

The partnership between the two agencies in this effort was so seamless that in their history of its implementation, Doug Smith and historian Gary Ferguson rarely indicate which agency the various actors were working for.[132] Once when I asked Smith about the level of participation of other agencies, he stated firmly that "the FWS was as enthusiastic about this as the NPS."[133]

The Power of an Informed Public

The coalition pursuing reintroduction of wolves at Yellowstone created conditions conducive to enlarging the sphere of conflict. They had redefined the image of wolves, provided powerful scientific and economic arguments, and fostered agency commitment, thereby gaining enthusiastic public support.

That support grew over time. Perhaps the earliest study of public opinion on the issue, a master's thesis at Colorado State University in 1977, found that three of four respondents were in favor of wolf reintroduction at Rocky Mountain National Park.[134] Another master's thesis, at the University of Montana in 1985, found support for wolf reintroduction among Yellowstone visitors by a margin of 3 to 1 and the belief that the "presence of wolves would improve the Yellowstone experience" by a margin of 6 to 1.[135] One is tempted to dismiss support among visitors as unrepresentative of the views of local residents or even the larger public. In fact, however, 1987 surveys by A. J. Bath showed that even residents of neighboring states were warming to the idea of reintroduction. He found more residents in favor of reintroduction than opposed in Montana (44 percent in favor, 40 percent opposed), Idaho (56–27), and even Wyoming (49–35).[136] Within those numbers are distinct battle lines. In one poll, Bath found 91 percent of the members of the Wyoming Stock Growers opposed to reintroduction; in contrast, 89 percent of the members of Defenders of Wildlife were in favor.[137]

The polls showed growing support for reintroduction in the 1990s. A 1992 national Defenders of Wildlife poll showed overwhelming support for reintroduction, with 97 percent of 35,000 responses in favor.[138] Such figures reflect national support from people, including city dwellers, who obviously did not face the potential dangers or property losses from predators that local residents did. Nevertheless, by 1990, another survey showed that Wyoming residents supported reintroduction by 3 to 1.[139] Political scientist John Freemuth surveyed Idaho residents about reintroduction in that state and found 72.4 percent in favor and only 22.1 percent opposed. Proponents of the reintroduction, such as Defenders of Wildlife president Roger Schlickeisen, could justifiably state that "time and again the public had overwhelmingly called for restoring wolves to their home in Yellowstone."[140]

How much did strong public support matter? When I asked John Varley, who was NPS chief of resources at the time of reintroduction, if it was fair to say that public opinion drove wolf reintroduction, he answered, "No question about that. Once we saw that a majority of the public in Wyoming, Montana, and Idaho supported reintroduction, then it was a matter of perseverance."[141] As Doug Smith stated emphatically, "the reversal of policy in this case came about from an informed public."[142]

Public support fostered official approval. Secretary of the Interior Babbitt, acting for the Clinton administration, approved the reintroduction plan in May 1994.[143] Proponents in Congress funded the program. Senator Burns of Montana later tried to amend the DOI appropriations bill for the second year in a way that would stop any more wolves from being brought to the park but had to settle for cutting one-third off the $600,000 budget.[144] The courts also endorsed wolf reintroduction. Livestock interests and their supporters (specifically the Farm Bureau Federations of Montana and Wyoming, the American Farm Bureau Federation, and the Mountain States Legal Foundation) filed suit to stop the program. On January 3, 1995, federal district court Judge William Downes ruled against the suit, writing that opponents of the program had offered only "fear and speculation" to support their position. He also cited some of the scientific evidence of wolves living in Minnesota without causing excessive trouble.[145]

Nonetheless, at the last minute, while the wolves to be transplanted were in transit, the Wyoming Farm Bureau asked the U.S. Court of Appeals in Denver for an emergency stay of the reintroduction. In spite of the fact that the wolves were going to be released only into acclimation pens from which they could be recollected if necessary, the court ordered the stay. Project scientists feared that they might have to euthanize the animals, which could not survive long in the transport boxes. Interior Department attorneys appealed the court's decision and, fortunately for the wolves, the court lifted the stay order after the animals had spent two full days in captivity.[146] Such legal battles would continue to fester until they resurfaced again a few years later, but the courts had shown an early inclination to side with pro-change forces.

The Predators Return

The best times to see wolves are shortly after sunset, when enough light remains for human vision, and shortly before sunrise, before the sun renders the animals relatively dormant. Knowing that and being too wired to sit around the campsite, I drove out of the campground well before sunset. Some people drive up and down the thirty-mile long road between Tower Junction and the northeast entrance to Yellowstone as if they're trawling on a deep-sea fishing boat, just looking for the traffic jams that signal a wildlife sighting. The road parallels the Lamar River through the valley, providing beautiful views of the wildflower-laden meadows and the surrounding snow-covered peaks. Even before the wolves were there, we used to come here if for no reason other than to enjoy the scenery.

I pulled over at a favorite spot and just sat, waiting for dusk. Within an hour, I had company. Earlier in the day, a nice couple in a pickup truck from Montana had stopped by my campsite to ask for wildlife-viewing advice, and realizing that they had no place to camp, I invited them to share my site. Now here they were, pulling into my turnout and asking the inevitable question, whether I had seen any wolves. Five minutes later, our mini-jam attracted another vehicle, an SUV from Colorado. A short conversation with them revealed some of those odd coincidences that make the world seem smaller. The woman in the SUV had a sister who at one time had been a babysitter for the Montana couple. Then the guy in the SUV suddenly looked at me and asked whether I had once given a talk at Colorado State University. I had. We all paused for a moment and then started laughing. Maybe meetings like ours aren't so surprising anymore in the Lamar Valley. For a lot of people, it has become the center of a certain kind of universe. We stood and talked and waited for the sun to go down and the wolves to come out.

A Success

The most dramatic reintroduction of a species into a U.S. wilderness occurred in early 1995. In January, program officials purchased fourteen wolves from Canadian trappers, moved them to Yellowstone, and placed them in acclimation pens. Program officials had tried to get wolves from areas like Yellowstone, which has a mountainous terrain and populations of elk and deer. The rumor among the livestock community was that cattle killers were the only kind of wolves that Canada sold, but the NPS and FWS purchased wolves from areas that had little livestock. They also bought packs and tried to keep them together, the idea being that wolves raising families would have less tendency to wander wide distances, including outside the park. When capturing the wolves in Canada, officials had attached tracking devices to individuals, released them, and then tracked the unwitting traitors back to their family units.[147] The large acclimation pens enabled a "soft release," whereby the transplanted wolves could bond and hopefully become accustomed to their "new" territory so that any tendencies to try to return from whence they came would be diminished. Even with those precautions, no guarantees were in effect when in March, the first wolves ventured from their pens and into the Yellowstone wilderness.

Ultimately, the wolves themselves made the reintroduction work. NPS and FWS officials took some precautions, such as fixing radio collars on the original wolves to monitor their behavior and releasing seventeen more wolves in 1996. But the wolves' survival was up to them. Indeed, when I visited the park in 1996, the NPS was beset by a range of issues that demanded attention, from a proposed gold mine on the northeastern border to nearly continuous public criticism of the park's fire control policies. The agency could devote only limited resources to monitoring and tracking the wolves. That was not that much of a problem because the animals took to their new surroundings as if they had always been there and knew exactly what to do. One remarkable event is illustrative. After wolf number 10, the father of the first group of pups born in the park, was killed by a local hunter after wandering south of park boundaries, program officials rounded up the mother and the helpless pups, put them back into the acclimation pen, and provided them with food for several months. Even when the pups got a bit older, program managers feared that the mother would have a hard time taking care of them by herself when they were released. Then, on the day they were to be released, in an event that no biologist has yet been able to explain, a bachelor wolf living miles away in another part of the park showed up outside the pen, just in time to form a new family unit.[148]

Inside the park, the wolves were thriving. Several things had gone even better than expected. First, the wolves bred successfully, despite the stress of relocation. By mid-1996, although nine wolves had died or been killed (one by official action after it was suspected of repeatedly killing sheep), twenty-three pups had been born in the park ecosystem.[149] The population continued to grow at a rate, roughly 17 percent a year, that exceeded even optimistic projections.[150] Second, the wolves did not try to migrate back to Canada, as some had feared. Third, as mentioned, they had not wandered out of the park to kill livestock very often, apparently happy with the wild game in the park. As wolf biologist Mech observed early on, "It is a wonderful start to a monumental conservation effort."[151]

In addition to the good biological news, the wolves had become instant favorites with the public. As figure 1-1 shows, visitation to Yellowstone remained high even when it was declining in other parks. Though all of it cannot be attributed to the wolves, they were a prime attraction. NPS ranger Rick McIntyre had become a celebrity in the Lamar Valley for leading about 40,000 visitors on "wolf walks" to see the animals. And Defenders proudly reported that wolves had become number one among the animals that tourists now wanted to see in Yellowstone.[152] Chief of resources Varley noted: "Until the mid-80s, people came to see the geothermal stuff, then wildlife became number one, then the grizzly bear became number one, [then] in 1995, wolves became number one."[153] Biologists began to document the distinctive attributes of each pack of wolves, and observers, both in the park and from afar, started following the exploits of the different packs and even of individual wolves.[154]

The conditions that had been conducive to formulating the policy change thus continued to support the ongoing implementation of the program. FWS officials concluded in 1996 that "the wolf recovery program is ahead of schedule, under budget, and is occurring with less conflict than predicted."[155] Similarly, when I talked with Yellowstone superintendent Finley about the wolf reintroduction in the late summer of 1996, he announced succinctly, "It's a success."[156]

Living with Them

As the wolves thrived in their new hunting grounds, they were oblivious to the challenges to the program that were arising. Not everyone was pleased with the apparent success of the reintroduction. Local resentment remained. One group of Montana ranchers printed "Wolf Management Team" t-shirts showing a wolf's head in a rifle sight next to the slogan "Shoot, Shovel, and Shut up." Chad McKittrick, the local hunter who had killed wolf number 10, became a bit of a celebrity for a while, showing up in a parade in Red Lodge, Montana, wearing a "Northern Rockies Wolf Reduction Project" t-shirt.[157] However, opposition among livestock owners was by no means uniform. For example, one fourth-generation rancher, Ben Cunningham, helped the reintroduction effort by hauling meat to the acclimation pens in 1995.[158] Still, when I asked Finley in 1996 if the early successes of the program had diminished opposition in neighboring communities, he answered emphatically, "No. In fact, support has dropped, largely because some wolves did kill sheep, so they can now say 'we told you so.'"[159] Not surprisingly, then, in the first five years of the program, poachers illegally killed seven wolves.[160]

Nevertheless, political support for the program remained strong. Funding, though occasionally challenged, has never ceased. After Senator Burns failed to stymie the project in 1995, others attempted to manipulate funding but with little success. In 1999, I asked Varley whether he ever worried about congressional cuts to the program. Roughly 17 percent of the park's $30 million budget in fiscal year 1999 had been designated for resource protection through programs including those for wolves; much higher percentages went to facility maintenance (40 percent) and visitor services (33 percent).[161] Nevertheless, in an answer that a bureaucrat is rarely able to give, he responded, "The funding for this program has not been cut; the NPS and the FWS have gotten everything they've asked for."[162]

The more serious challenge came in the courts, testing the strength of the support for the reintroduction. The courts ruled again on the wolf program in 1997. Late that year, Judge Downes, the judge who had ruled in 1995, issued an opinion on the goal of establishing experimental populations of species. Downes based his ruling on the language in the ESA that allows experimental populations only when, according to Section 1539 (j)(1), they are "wholly separate geographically from nonexperimental populations of the same species." The plaintiffs claimed that naturally occurring wolves had been sighted in the same geographical area as the reintroduced wolves; how, then, could the two sets of wolves be treated differently? Somewhat surprisingly, in addition to the Wyoming Farm Bureau Federation, the plaintiffs included the Audubon Society and the Sierra Club Legal Defense Fund, which wanted full endangered status for the wolves already living in Idaho and argued that wolves were returning to different areas in the Rockies on their own.[163] Judge Downes assessed the Idaho and Yellowstone wolves together and ordered the removal of the experimental wolves. Downes stated that "Congress did not intend to allow reduction of protections to existing natural populations in whole or in part."[164] Expressing reluctance in his decision, given the impressive results of the program, he issued a stay to allow appeals.[165] Downes stated that his decision was "not intended to serve as a license to euthanize wolves."[166] Still, the remove order could have been a death sentence for the wolves as no alternative facilities or sites were available.

Wolf supporters and other observers rallied in response. All the components of the strong coalition that had pushed the initial reintroduction were again evident. The National Parks Conservation Association warned that since removal would be difficult, "the park could be forced to kill hundreds of wolves."[167] Writer Thomas McNamee castigated the complicit (and to some traitorous) environmental groups for their "legalistic absolutism," which may "have helped destroy one of the conservation movement's greatest triumphs."[168] The *New York Times* called the ruling "a sad encouragement to the mistaken defensiveness of livestock groups who purport to represent the interests of most ranchers."[169] In early 1998, shortly after the Downes ruling, a nationwide poll conducted by the National Wildlife Federation showed that even a majority of respondents in neighboring states opposed the remove order.[170] Secretary of the Interior Babbitt promised to "fight with everything I have to keep the wolves in Yellowstone where they belong."[171]

Whether or not the public outcry made the difference, the courts again endorsed the restoration effort. In January 2000, the 10th U.S. Circuit Court of Appeals ruled that the wolves could stay in the park. The judges cited the ultimate goal of protecting species and endorsed the flexibility allowed in the ESA, in effect allowing experimental species. The appeals court stated that a limit to that flexibility such as that imposed by Downes "ignores biological reality and misconstrues the larger purpose" of the law.[172] The fact that this ruling came from the 10th Circuit Court was important also because it was different from the circuit court that had upheld the conviction of Chad McKittrick for killing wolf number 10. After McKittrick appealed, he lost in the 9th Circuit Court of Appeals; the U.S. Supreme Court later refused to hear the case. Thus, two different appellate courts had ruled in favor of the restoration project. Secretary of the Interior Babbitt applauded the

NPS PHOTO BY JIM PEACO

The reintroduced wolves have affected nearly every other species in the park. They rarely attack bison, but interactions such as the one in the photo do occur.

denial of the remove order, describing it as "a ringing endorsement to our wolf reintroduction program."[173]

Over time, many local residents, even ranchers, grew more resigned to the presence of the wolves. A major reason was that the number of attacks was quite low, much lower even than anticipated in the EIS. In the first four years of the program, the FWS estimated that wolves in the northern Rockies had killed fewer than 100 cattle and 200 sheep. The agency could in fact confirm only twelve cattle and 100 sheep kills due to wolves.[174] The disparity between the numbers is due to the fact that predation is difficult to prove, especially if carcasses have decayed by the time they are discovered. To give some perspective on these numbers, Montana livestock ranchers alone lose thousands of sheep and cattle each year to all causes. Admittedly, any livestock losses can be problematic for ranchers, especially those with smaller herds.[175] Further, obtaining the compensation promised for wolf kills is not easy. Ranchers must report kills to the NPS within twenty-four hours and must present evidence that the death was due to wolves and that the rancher had observed proper protection procedures. As one rancher in Montana told me, proof is difficult if there is almost nothing left of the dead animal. Nevertheless, by mid-1999, the Defenders of Wildlife had compensated ranchers with nearly $70,000 for wolf-caused losses.[176] While one could not claim that the program had converted the livestock community to wolf fans, John Varley did say that "many ranchers have confided their support privately, and many now say, let's live with them."[177]

By mid-1999, approximately 110 adult wolves and as many as sixty pups in ten different packs roamed the park. In Idaho, the companion reintroduction had resulted in similar numbers of wolves. In addition, wolves migrating south from Canada had formed eight packs with nearly sixty adults and fifty pups in northwestern Montana. In at least two instances, wolves from two different populations had mated, thus suggesting the potential for a genetically diversified population through the northern Rockies.[178]

Because of Those Damn Wolves

While taking a cab to the Billings, Montana, airport several years after reintroduction, I noticed a lot of construction activity. I asked the driver, a surly older guy who had lived there for decades, about the apparent local boom, and he replied, "I'll tell you why the economy is so good, it's because of those damn wolves." I wasn't sure whether he recognized the irony of simultaneously damning and praising the wolves.

While their contributions have not won over all the local residents, the animals have been extremely successful on several fronts, including the stimulation of the local economy. NPS management assistant John

Sacklin told me in 2005, "The wolf reintroduction has been successful by any standard."[179] Its success was supported by the sort of scientific and economic evidence that helped compel the reintroduction in the first place.

Scientific Evidence. In fact, a 2005 assessment by a half-dozen noted biologists compared the progress after ten years to the predictions in the EIS (summarized in table 2-2) and found that the project had equaled or exceeded the most optimistic expectations.[180] One decade after reintroduction, more than a dozen wolf packs with a stable population of just over 300 lived in the Greater Yellowstone area, about 200 of them in the park itself.[181] By 2008, the wolf population in Wyoming, Montana, and Idaho together totaled more than 1,500 and was growing at the rate of 24 percent a year.[182] The Yellowstone success also inspired efforts to bring wolves to other wilderness areas, including Olympic National Park.[183] The wolf population has remained robust in spite of the predictable poaching and the less predictable damage from a disease of dogs known as parvovirus.[184] The EIS had predicted that recovery would be attained in eight years, and the program met biological recovery criteria by 2002.[185] One analysis called it "the most successful recovery effort in the history of the Endangered Species Act."[186]

Success in ecological terms was substantial and apparent early on, as the wolves quickly and dramatically restored natural cycles that had not existed since they were eradicated from Yellowstone.[187] However, the contribution of wolves to the needed decline in the size of the elk population should not be overstated. One three-year study found that more than half of the elk deaths in the park were attributable to grizzly bears and less than a quarter to wolves.[188] But, as the EIS anticipated, elk have been the main food source for wolves, constituting 92 percent of their kills in winter months.[189] One impact of the wolves on the elk seems to have been to force the elk to become more vigilant and to move to higher ground, thereby benefiting the plant species in river bottoms, such as aspen, that were suffering from the bulging elk population. Again, some scientists argue that wolves are just part of the aspen recovery, vegetation throughout the park having benefited from both the natural fires in 1988 and some prescribed burning.[190] Nonetheless, while willow recovery has been variable throughout the park, it has been

stronger where wolves are more active. In those riverine areas, beaver have rebounded, with the result that their dams are again shaping rivers in ways that assist other wildlife. In addition, the wolves killed roughly half the coyotes in the park (although their numbers have recovered somewhat in recent years), thereby contributing significantly to huge growth in the populations of voles, ground squirrels, and gophers, which are major parts of the coyote diet. With fewer coyotes, more rodents are available to other predators, such as foxes and hawks. Even grizzly bears have benefited. Scientists reported numerous cases of grizzlies feeding on wolf kills. Biologists refer to such effects as a trophic cascade, and in Yellowstone the benefits have been substantial.[191]

Many scientists are euphoric about the program. A leading biologist with Yellowstone Ecosystem Studies, Robert Crabtree, summarized the wolf reintroduction by declaring, "Wolves are causing an explosion in species diversity."[192] Wolf project leader Smith, also a biologist, agreed: "I think what the wolves are doing to Yellowstone is providing balance."[193] Today, few doubt that the restoration of wolves within the ecosystem has been beneficial to species ranging from eagles to aspen.[194]

The success of the wolf reintroduction in the greater Yellowstone ecosystem has not translated to success for all species. During the same period, the same agencies struggled with other wildlife programs, some inevitably affected by the wolves. In 1999, I talked with Chuck Schwartz, then head of the Interagency Grizzly Bear Study Team, to compare efforts to sustain populations of wolves and grizzlies. He answered, "The incredible success of the wolf recovery is a no-brainer compared to the grizzlies."[195] In other words, determining whether grizzly recovery had been successful was more complicated than determining whether the wolf program was a success. The population of grizzlies somehow has increased in spite of daunting threats and challenges, but many observers remain wary, given the loss of key habitat and the endangerment of important food sources for the bears, such as whitebark pine nuts and native trout species. In sharp contrast to the wolves, bison have suffered tragic losses in recent years. Concerns over the disease brucellosis have led state and federal officials to kill literally hundreds of bison that wander outside of the park, looking for food, in winter.[196] Many arguments over the bison have centered on the intrusiveness of

snowmobiles in Yellowstone and whether snowmobile trails make it easier for bison to roam outside of the park to their doom.[197] Even as late as 2008, federal and state authorities were struggling with an adequate response to the bison situation.[198] Other species in the park face significant threats from invasive species such as lake trout. The Greater Yellowstone Coordinating Committee, consisting of the managers of the different units of federal lands in the area, are charged with addressing these issues through an approach that takes the entire ecosystem into account, but that has proven to be challenging.[199]

Economic Evidence. Economic benefits to the larger community from wolf recovery became apparent very early in the program with the increase in tourists, particularly wolf-watchers. The Lamar Valley may well have become "the best place for wolf-watching . . . in the entire world."[200] Just five years into the program, the NPS estimated that at least 15,000 people a year were coming to see wolves in Yellowstone;[201] today, more than 150,000 visitors have done so. Putting a dollar figure on such experiences is difficult, but analyses show some impressive results. One thorough study of nearly 3,000 survey responses between late 2004 and early 2006 concluded that approximately 94,000 people came to the park from outside the GYE each year to see or hear wolves; they spend an average of $375 per person, for a total annual benefit of more than $35 million.[202] The estimates vary, but other studies credit the wolves with economic benefits of between $20 and $28 million a year.[203]

On the other side of the ledger, the wolf project does generate some actual costs as well as some opportunity costs. Costs of implementing the program have been consistent with projections in the EIS. Actual costs to reintroduce the wolves totaled about $870,000 for Idaho and Yellowstone combined, while costs to monitor and manage the program averaged about $1.5 million a year.[204] In addition, wolves do kill some livestock. Biologists estimate that confirmed wolf kills in the GYE between 2000 and 2003 averaged twenty-seven cattle and seventy-nine sheep a year.[205] Although those numbers are small compared with those for livestock losses from other causes, admittedly the livestock industry may also have to absorb costs of increased fencing and vigilance. Further, predation is likely to increase with the increasing number of wolves. In 2004 for instance, Defenders of Wildlife paid ranchers

$136,838 for stock killed by wolves.[206] Economists estimate that even if the Defenders' numbers understate livestock losses by half due to verification issues, recent direct losses totaled less than $130,000 a year.[207] Still, any losses to ranchers can be tough, especially in hard economic times. Thus, Defenders and other groups have recently begun to assist ranchers more actively in protecting their livestock from wolves. Examples include providing range riders to guard cattle herds and solar-powered electric fences to protect sheep.[208]

There also may be some opportunity costs of reintroduction. Hunters and outfitters argue that the thriving wolf population has reduced elk herds to the point of substantial recreation losses. Indeed, Doug Smith now considers hunters the "most vehemently anti-wolf people."[209] The facts do not support the hunters' complaints, however. Elk numbers have recently climbed to roughly 350,000 in the area, and hunters killed more elk in 2005 than they did in the years before wolf reintroduction.[210] The 2005 biological assessment showed that the average annual harvest of elk by hunters between 1995 and 2004 (after reintroduction) was actually higher than the average between 1976 and 1994 (before reintroduction).[211]

Other potentially relevant costs and benefits are harder to attribute directly to the wolves. The benefits to the larger U.S. and international publics of just knowing that the wolves are there in the premier U.S. ecosystem defy calculation. Whether the presence of wolves has had any impact, either positive or negative, on relocation decisions, the region's population grew by 12 percent in the twenty-five-county area surrounding Yellowstone between 1995 and 2005.[212] One has only to wander in to any gift store in the region to see that local merchants have adopted the wolves, at least for the sake of selling souvenirs. And as the comment of the cab driver quoted earlier suggests, any argument that wolf recovery has slowed regional economic growth is hard to support.

So, what is the bottom line on the value of reintroduction? Two recent studies are explicit. A group of economists estimated in 2008 that weighing just the direct economic benefits of tourism against loss of livestock and of hunting opportunities produces a net economic benefit of well over $30 million a year.[213] In a separate study, a ten-year scientific assessment estimated that actual and opportunity costs would

have to increase more than fourfold from current levels even to approach the lowest estimates of total benefits from the program.[214]

Environmental Poster Child

The wolf reintroduction has gone so well that, somewhat ironically, the wolves are now threatened by their own success. Indeed, virtually all the conditions for strong public support that were evident in the early years of the program remain intact. The scientific and economic studies cited above support the original predictions of benefits, and agency officials remain committed to the policy. Yet some political actors remain hostile to the program. As NPS management assistant Sacklin said, "No amount of good science will stop a politician."[215] Thus, ecological and economic benefits alone are not sufficient to protect the program. Fortunately for the wolves, their image has remained favorable among the larger public. As wolf project leader Smith says, "Wolves have become the environmental poster child for the twenty-first century."[216]

That image has been crucial in recent developments. In the mid-2000s, the George W. Bush administration, acting through neighboring state governments, attempted to reduce the protections afforded to the wolves. As mentioned earlier, the wolves achieved the recovery goal of at least 300 wolves in the Yellowstone region and thirty breeding pairs in 2002. In early 2003, the Bush administration proposed downlisting the gray wolf from endangered to threatened in most of the lower forty-eight states, a move that would transfer responsibility for protection of the wolves to state governments when wolves are found within state jurisdictions.

Delisting was not supposed to occur until the federal government had approved the management plans of the three neighboring states. Thus, wolf proponents, wary of state intentions, filed a lawsuit to stop the action. On February 1, 2004, U.S. District Court Judge Robert Jones ruled that the downlisting was based on political expediency rather than science. In January 2005, the administration issued new rules stating that a landowner could kill a gray wolf in Idaho and Montana without waiting for physical evidence of an attack if the landowner believed that the wolf was in the process of killing livestock. One month later, Jones ruled against those provisions, stating that the rules did not take

NPS PHOTO BY DOUG SMITH

Wolves and grizzly bears are both loved and feared by many people. Today, they are powerful reminders that Yellowstone was established and restored to be a natural, wild ecosystem.

into account enough factors in evaluating the wolf's status.[217] Delisting therefore was in temporary limbo. When I visited the park in 2005, John Sacklin and planner Mike Yochim both concluded that while acceptance of the wolf program in neighboring states had increased to some extent, for the most part state officials were just "biding their time," waiting for delisting.[218]

In 2007 and 2008, the administration moved systematically toward delisting in spite of the fact that in 2007, state authorities in Idaho and Wyoming promised liberal hunting rules if and when federal protection was lifted.[219] In Idaho, a hunter could buy a wolf permit for just $26.50. Even without such a permit, residents were free to kill a wolf just for "annoying" domestic animals.[220] Idaho Governor Butch Otter said that he would support hunts to kill all but 100 wolves in Idaho and looked forward to shooting one himself.[221] Wolves in Wyoming outside of Yellowstone would be labeled predators and subject to hunting, no permits required and no limits imposed. In response, environmental groups, notably Defenders of Wildlife and the Natural Resources Defense Council, mobilized opposition to the delisting, eliciting tens of thousands of written comments against the proposed action.[222] Huge majorities, even among residents in the counties around the park, opposed delisting while the Wyoming plan was in effect. For instance, 95 percent of comments from Teton County, near Jackson Hole, were critical of the plan and the likelihood that more wolves would be killed than necessary.[223]

The states responded but only to a limited degree. Wyoming state officials modified their plan so that wolves in the Yellowstone area would be classified as trophy game animals to be killed only in licensed and controlled hunts. Ultimately, policymakers in each of the three states committed to maintaining wolf populations within their borders of at least fifteen breeding pairs and 150 wolves.[224] Those modifications enabled the Bush administration to delist the wolf in early 2008. In February, Deputy Secretary of the Interior Lynn Scarlett announced the decision by stating emphatically, "Gray wolves in the northern Rocky Mountains are thriving and no longer need protection."[225]

Many scientists and members of environmental groups worried that the numbers did not in fact warrant removal and feared a pending slaughter in neighboring states. Environmental groups filed a lawsuit to reverse the decision.[226] As if ordained to reinforce wolf lovers' fears of hunting, on the day after delisting a hunter in Wyoming killed three wolves near an elk feeding ground. In a classic illustration of the power of image, the hunter not only created a front-page story in the Mountain West but also an instant martyr. One of the animals killed was perhaps the most photographed wolf in the country, a seven-year-old radio-collared male that fans referred to as "Limpy."[227] Limpy earned his nickname after roaming more than 200 miles to Utah, where he was injured in a coyote trap. After being re-released into Grand Teton National Park, he limped home sixty-five miles to his family in Yellowstone. But he did not survive delisting. Within a month of delisting, hunters had killed more than twenty wolves in the Rockies, sixteen alone in Wyoming, a rate that caused significant concern among activists and scientists.[228] Within two months of delisting, hunters had killed sixty-nine wolves.

Legal and popular support for the Yellowstone wolves remained strong, however. In July 2008, Federal District Court Judge Donald Molloy ordered the Bush administration to restore endangered species protection for the wolves. The judge ruled that the administration, acting through the Fish and Wildlife Service, had acted arbitrarily in delisting. Thus, the states could no longer legally act on their own management plans and allow continued killing of the wolves. Even the *Casper Star Tribune* criticized Wyoming state officials for their "stubborn resistance" to some regulations on the hunting of wolves.[229] Michael Scott of the Greater Yellowstone Coalition encouraged the states to "begin fixing the flaws in state-management plans" rather than prolong the court battles.[230]

Doug Smith, probably the most knowledgeable person with regard to the Yellowstone wolves, was realistic about delisting and the possible consequences. In an interview in late 2008, Smith addressed several issues.[231] First, delisting will occur. As Smith says, delisting has to occur if for no other reason than that he and other federal officials promised that it would once the wolf population reached certain levels, goals that have now been attained. Therefore, Smith has testified in favor of delisting.[232] A second and related issue is that if delisting is not done properly, hunting in neighboring states will draw down wolf numbers substantially. Wolves tend to move from high-density to low-density areas, so if hunting reduces numbers in neighboring states and wolves move out of high-density Yellowstone and into border areas, indiscriminate killing could produce an "indirect harvest on wolves in the park." Thus, Smith and others argue that more conscientious state policies than exist now, particularly in Wyoming, are essential. A third issue involves genetic diversity. One is tempted to say that even when the wolves are delisted, at least the wolves in Yellowstone will remain protected. However, others worry that an isolated population, even in an ecosystem as large as Yellowstone's, could eventually result in a loss of genetic diversity.[233] Smith and other wolf biologists are less concerned about that possibility than many environmentalists. He argues that, although a few cases of Yellowstone wolves interacting with Idaho wolves have been documented, little natural "connectivity" to other wolf populations now exists anyway because of the lack of "avenues for wolves" into the park. In addition, genetic diversity within the park's wolves is high and at least a century will pass before any diversity problems are manifest. If the potential does increase, as Smith says, nothing says that connectivity to other wolf populations has to be natural, so the NPS and FWS could bring in more outside wolves to expand the gene pool in the future.

In late 2008, the Bush administration moved again to delist the wolves. The FWS opened its proposal for delisting for public comment, stating that it hoped to get comments related to the recent court actions. One possibility was to delist in all locations except Wyoming.[234] After coming into office, the Obama administration officially endorsed delisting. Members of the coalition that pushed for and supported wolf reintroduction mobilized again.

The Times They Are A-Changing

Early in the morning after a short night at the Pebble Creek campground, I sat on a hillside in the western end of the Lamar Valley and witnessed the most remarkable thing that I've ever seen in any natural setting. The night in the valley had already been successful for wolf watchers. We had spotted a few wolves the evening before but caught only glimpses as they stealthily hustled out to hunt in the darkening meadows. We certainly heard them, though, calling to one another from seemingly miles away, their howls a sound that once heard is not likely to be forgotten. And earlier in the morning, several of us had watched some wolves chase a few coyotes away from an elk kill in a small meadow on the other side of the river from the campground.

The scene that now lay before me, half a mile away in an immense meadow formed by a gradual bend of the Lamar River, was stunning. Ignoring the nearby herds of pronghorn antelope and bison, a pack of wolves had settled down to eat an elk that they had killed the previous night. Suddenly, a grizzly bear mother and her two cubs entered the meadow from the far western end, sniffing the air briefly and then making a beeline toward the apparently aromatic elk carcass at least a quarter mile away. The wolves, ever vigilant, saw the bears coming, and moved to defend their breakfast. Like two trains heading for a collision on the same track, the five wolves and the three grizzlies approached each other, both moving quickly at first and then slowing slightly just before the crash. For the next two hours, we watched from the hillside, using binoculars or scopes or only our eyes when necessary, witnesses to a struggle that was both primordial and awe-inspiring. The wolves, not surprisingly, attempted to use teamwork, luring one of the cubs away from its mother and then trying to surround the outnumbered bears. The cubs, most likely in their second year and therefore of decent size, were like aggressive teenagers; they were game for at least a brief skirmish but then wisely scampered back to their mother, who would stand on her hind legs and swat away the pursuing wolves. As the battle moved seemingly inexorably toward the carcass, others on the hillside, obviously wolf lovers, began to criticize the bears as poachers. Personally, I admired their determination and realized that if a hungry grizzly sees food, nothing short of death will stop it. Indeed, eventually the bears reached the carcass and took turns standing guard while the others ate until they had all had their fill. The wolves prowled nearby, resigned to but not happy with the prospect of leftovers.

I would not, however, say that the bears were the biggest victors that day. We were. We were privileged to witness such an event, beneficiaries of the efforts of people who believed at some point, as in the Bob Dylan song that provides the title for this concluding section, that "the times they are a-changing."[235] The scene we saw that day in Yellowstone was possible only because of the dramatic repair of a policy that had long diminished this amazing place.

The Merced River flows through Yosemite Valley, framed at the western end by El Capitan and the Cathedral Rocks.

3

REDUCING AUTOMOBILE IMPACTS AT YOSEMITE

Increasing automobile traffic is the single greatest threat to enjoyment of the natural and scenic qualities of Yosemite.

—YOSEMITE GENERAL MANAGEMENT PLAN, 1980

The plan is a big hammer. What we're looking for now is not a big hammer but a lot of small fixes.

—LINDA DAHL, CHIEF OF PLANNING, YOSEMITE NATIONAL PARK, 2008

Every day in Yosemite invigorates the senses. On a quiet day, you can hear the sound of the cold, clear water bubbling in the creeks and the Merced River crashing over the rocks. The scent of the ponderosa pines is everywhere, especially noticeable in the early morning when the trees are just being awakened by the sun. As for the sights, Teddy Roosevelt once called the mile-wide, seven-mile-long Yosemite Valley, with its stunning combination of granite cliffs, waterfalls, and meadows, "the most beautiful place in the world."[1] Many other people agree. Having worked there and visited in all four seasons, I knew it to be true. But I also knew that things had changed over time. For most visitors to Yosemite in the latter part of the twentieth century, the assault on the senses was somewhat different. The soothing sound of the Merced was sometimes just audible over the roar of automobile engines. The sweet scent of the pines was occasionally overwhelmed by the smell of gasoline and exhaust fumes. Sightlines were often obstructed by traffic jams. So, as we drove toward the valley one July day in the fire-stricken summer of 2008, I had to admit to being a bit pensive. My apprehension was not due to the fires; I was fortunate in coming after they had subsided. It arose from the evidence of human impact. Had the managers of Yosemite actually made the changes promised in the 1980 general management plan?

Perhaps nowhere else in the park system—or, for that matter, among other protected areas throughout the world—is the tension between access and protection, visitor use and preservation of natural conditions, as vivid as in Yosemite. For decades now, the managers of Yosemite have recognized the threat from automobiles mentioned in the opening quote and have promised to address it. They have made some progress, but as the quote from chief planner Linda Dahl suggests, they are focusing more on what is possible than on what is optimal. This chapter focuses on efforts to reduce automobile traffic, but those efforts necessarily include other aspects of development in the valley.

Putting Nature's Choicest Treasures at Risk

Even the earliest visitors to the Yosemite area must have recognized the valley as a special place. Although there are no written records of their reactions, Indians lived there for centuries. Whites began exploring the area in the mid-nineteenth century. The most famous early visitor, John Muir, described Yosemite Valley as a "temple" with "striking and sublime features on the grandest scale" and wrote that "into this one mountain mansion Nature had gathered her choicest treasures."[2] Like any treasure, though, those in Yosemite require protection, and that has not always been forthcoming.

Construction of Roads and Paths

The tension between protection and use in Yosemite has been evident throughout its history, a tension that is almost inevitable if the place is to be accessible. Even before Muir's first visit in 1868 and the establishment of Yellowstone in 1872, some policymakers realized that Yosemite warranted protection. Settlers had already staked claims and built some primitive structures for lodging.[3] In 1864, Congress deeded the Yosemite Valley and the Mariposa Grove, an area of Giant Sequoia trees thirty-five miles south of the valley, to the state of California under the condition that the lands "be held for public use, resort, and recreation . . . inalienable for all time."[4] The bill passed with very little debate or even much public notice.[5] Yet, even then, people were trying to figure out how to use, in particular how to access, Yosemite for private benefit. By the 1870s, many curious people wanted to visit, particularly the valley that so inspired anyone who saw it. But just getting to the park was a monumental task, involving taking a boat to Stockton, a sixteen-hour stagecoach ride to Coulterville, and then a thirty-seven-hour ride by horse and mule.[6] In an attempt to make the journey smoother and more profitable, private citizens built toll roads in the 1870s from three small towns to the west of the Valley (Coulterville, Mariposa, and Oak Flat). Once the roads were constructed, the number of visitors increased significantly enough to warrant extending the stagecoach lines.

When the federal government established the national park in 1890, it left the valley under state jurisdiction. Some in the state, including the first chairman of the board of commissioners, Frederick Law Olmstead, recognized the great beauty of the park and the associated need to try to protect it "as exactly as is possible."[7] Others were much less careful. Reports of state mismanagement, including a Department of the Interior study citing "great destruction," increased the pressure to add the valley to the national park holdings. The state had allowed timber cutting, agricultural development, stringing of barbed wire fences, and grazing by hoofed animals in most of the valley. John Muir and others pushed for expansion of the park and greater protection from dangers such as "hoofed locusts" (sheep).[8] In 1905 and 1906, the state therefore "regranted" the valley back to the federal government.[9]

At that point, the valley became subject to the requirement for protection in the 1890 legislation. In it, Congress mandated that managers "shall provide for the preservation from injury of all timber, mineral deposits, natural curiosities, or wonders within said reservation, and their retention in their natural condition." The same section of the legislation, however, also allowed for "the erection of buildings for the accommodation of visitors" and "the construction of roads and paths."[10] Thus, even while calling for preservation, the door had been officially opened to more access and use. One of the more significant

The state of California opened the All-Year Highway into Yosemite in 1927 to allow automobile access to Yosemite Valley in all four seasons. The road still carries thousands of vehicles into the valley each year.

developments occurred in 1907 with the extension of the Yosemite Valley Railroad from Mariposa to El Portal, the gateway community just west of the valley. Visitors could thus ride a train to El Portal and then catch a coach into the valley.

In the meantime, however, the valley had already experienced a new arrival. With symbolic timing, Oliver Lippincott drove his steam-powered automobile over the rough road from El Portal into the valley in 1900, the start of the new century. Other automobiles followed. Fearing the consequences of noise and possible accidents, acting park superintendent H. C. Benson banned cars in 1907.[11] In a foreshadowing of the controversy to come, the ban spurred protests from motorists and the California Automobile Association. Secretary of the Interior Franklin Lane reversed the decision in 1913, stating that "this form of transportation has come to stay."[12] Two years later, Assistant Secretary of the Interior Stephen Mather and some associates purchased an abandoned roadway over Tioga Pass, the most accessible route into Yosemite from the east; they deeded it to the park and called on the Bureau of Public Roads to construct a route into the park from that direction. In agreeing to do so, these national policymakers initiated a long-standing commitment to making Yosemite accessible by car. They had at least one prescient warning of the inherent dangers. Early in the twentieth century, British ambassador Lord James Bryce said of Yosemite Valley, "It would be hard to find anywhere scenery more perfect," but then he added ominously, "If Adam had known what harm the serpent was going to work, he would have tried to prevent him from finding lodging in Eden; and if you were to realize what the result of the automobile will be in that wonderful, incomparable Valley, you will keep it out."[13] Bryce's warning was ignored.

With the automobile came visitors and ultimately more roads. By 1914, Yosemite was hosting just over 15,000 visitors a year. In 1915, the number more than doubled, to 33,452, more visitors than at any park except Hot Springs (115,000), Yellowstone (52,000), and Mount Rainier (35,000).[14] The demand to see Yosemite was high enough to require more roadwork. In addition, as visitor numbers increased, so did the services provided. That meant gas stations, lodging, places to eat, parking spaces, and other facilities. The origins of the development in the park were modest. Camp Curry, still prominent in the valley today, grew out of a small permanent camp established by a couple of transplanted Indiana schoolteachers in 1898. The Curry operation became Yosemite Park and Curry Company (YPCC) in 1925. The YPCC built gas stations, restaurants, hotels, and other establishments in the valley over the next several decades. In the 1920s, the YPCC used the lure of winter sports, such as a skating rink at Curry Camp and skiing at Badger Pass, to encourage making the park a popular destination even in what had been the off-season.[15] In 1924, the California Highway Commission began building what was termed the All-Year Highway along the Merced River into the Valley. This road, now California Highway 140 from Merced and through El Portal, made the park accessible to visitors in all seasons.

The historic, multi-day trek by coach and horse into the valley had been replaced by a multi-hour drive from population centers such as the Bay Area. In 1926, the NPS signed an agreement with the Bureau of

Public Roads to upgrade roads within the national parks. The bureau pursued its responsibility aggressively in Yosemite, paving valley roads and constructing the routes from Wawona and Big Oak Flat. Bureau engineers, often with the help of the Civilian Conservation Corps, were at least somewhat conscientious with their work, building tunnels and bridges to minimize scarring and retaining as much of the landscape as possible. As historian Richard Quin said, "the roads were designed not only to carry heavy traffic loads, but to appear to be a part of their settings."[16] Whether or not their appearance was convincing, soon enough the roads were indeed carrying heavy traffic.

Visitation and the accompanying development continued to grow. Figure 1-1 in chapter 1 displays visitor growth over the last century. By 1954, more than a million people were visiting Yosemite annually, most of them spending at least some of their time in the valley. The increase and the accompanying crowds were noticeable. In 1966, with Yosemite visitation near 2 million, the *Wall Street Journal* ran a front-page story under the headline "Ah Wilderness: Severe Overcrowding Brings Ills of the City to Scenic Yosemite." The article claimed that on an average summer day, the valley hosted three times as many people per square mile as Los Angeles County and more per square mile than Chicago's Cook County. The crowds translated to "a soaring crime rate, incredible traffic snarls, juvenile rowdyism, and even smog."[17] By 1970, more than 2.2 million visitors were coming each year, inevitably bringing with them some of the behavior that many were trying to leave behind. The 1970 Fourth of July riots in Stoneman Meadow, for example, involved rock throwing, fights, and attacks on rangers throughout the day and evening.

Development efforts intensified in the 1970s. In 1968 the Park Service had ended one of the most questionable activities, creating a firefall by dropping burning wood off a cliff (shown in the movie *The Caine Mutiny*), but other activities were just as intrusive. Convinced that people needed a more user-friendly wilderness, the Music Corporation of America (which bought the YPCC in 1973) and its subsidiary, Universal Studios, pursued such activities as filming a television series (*Sierra*) in the valley and proposing a tramway to Glacier Point. In the filming of *Sierra,* the crews occasionally stopped traffic and even painted some rocks for one sequence.[18]

The concessionaires continued to provide more services as more and more visitors came to the park. Visitation increased to 4 million by 1990. By the latter part of the twentieth century, Yosemite Valley had become "the symbol of the national park ideal both at its finest and at its worst."[19] Critics began to demand that the NPS do something about development and, in particular, traffic. One pointed comment came from Edward Abbey, the man the *New York Times* called "the next literary guru to the nation's campus readers," who wrote in 1977 that riots had no place in the park and then added, "I can think of other things Yosemite Valley is not the proper place for. It is not the proper place for paved roads and motor traffic in any form."[20]

The Plan to Remove All Private Vehicles

NPS officials have long been aware of the problems resulting from traffic and development in the valley, but doing something about them has been difficult. Throughout the history of efforts to change policies in Yosemite, how to define the issue has never been entirely clear. Whereas proponents see changes as restoring natural conditions, opponents brand them restrictions on freedom and access.

Over time, the NPS has contributed to confusing the framing of the issue by sending its own mixed messages. Efforts to reduce commercialism in the valley date at least to 1949 (see table 3-1). The master plan for the park written in that year proposed moving some overnight accommodations out of the valley to the Big Meadows area just to the west.[21] The proposal was never put into effect, but officials renewed it in the 1960s. Then, as part of the Mission 66 agenda, NPS director Conrad Wirth tried again to transfer accommodations to Big Meadows and other facilities to El Portal, leaving much of Yosemite Valley for day use only. Wirth had to settle, as he said in his own book, for moving the concessionaires "from the center to one side of Yosemite Valley."[22] But Wirth also did his part to encourage development in the valley, awarding the same concessionaires a thirty-year contract for all commercial operations in the park in return for a minimal franchise fee of .75 to 1

Table 3-1. *Timeline of Actions Related to Policy Changes at Yosemite*

Year	*Policy action*
1890	Yosemite National Park is established.
1949	Master plan proposes moving facilities outside of park.
1980	General management plan proposes eliminating cars from valley.
1987	Yosemite Valley/El Portal Comprehensive Design outlines costs and options for changes in the valley.
1991	Scoping process begins for Yosemite Valley Plan.
2000	Draft EIS for management and restoration of Merced River Valley.
2000	Comment period on draft EIS between April and July.
2000	Yosemite Valley Plan outlines changes proposed for the valley.
2006	U.S. district court rules against Merced River Plan.
2008	Circuit court upholds district court ruling on Merced River Plan.

Source: U.S. National Park Service, *Yosemite Valley Plan: Final Supplemental Environmental Impact Statement* (Washington: Department of the Interior, 2000).

percent of gross receipts.[23] In 1969, NPS officials again developed recommendations to move housing and transportation to a staging area but did not act on the idea.

The agency did take some significant steps in the 1970s toward reducing traffic and development, but those also met with resistance. In 1970, the NPS closed the eastern third of the valley to automobiles, mandating that the area be served only by trams and shuttle buses powered by propane. Further, the agency even considered a plan to remove all automobiles from the entire valley.[24] A committee of NPS officials and interested participants, including the president of the Sierra Club, drafted a plan that would move many facilities out of the valley. The NPS also began making changes in the park, for example, cutting campground use and making roads one-way. Those seeing such changes as restrictions on use began drawing battle lines. Concessionaires, anticipating declining sales as a result of the changes, appealed to political authorities. For example, Edward Hardy, chief operating officer of the YPCC, made his own suggestion for park planning to park superintendent Les Arnberger, stating that "planning should focus on alternate travel options, such as the aerial tramway to Glacier Point and increased parking within the valley."[25]

The new draft of the plan released by NPS officials did in fact call for expanding accommodations in the valley. Not only did it not include many of the proposals for removing facilities, but as the NPS director admitted, it did not deal with automobiles at all. That inspired environmentalists, such as Connie Parish of the Friends of the Earth, to express their "dismay that resolution of the parks' main problem—cars—was put off."[26] Critics even suggested that the plan had been written by the concessionaires, inspiring congressional hearings and investigations that, although they addressed problems found throughout the park system, often focused on Yosemite as one of the worse cases.[27] Members of Congress wondered aloud whether the NPS or the concessionaires were running the parks and demanded that the public be involved to a much greater extent in future planning efforts. In response, Assistant Secretary of the Interior Nat Reed rejected the draft plan and ordered a new one to be prepared with public input. But by then, as one historian wrote, "the Park Service looked like a weak sister, an outfit easy to manipulate, especially at Yosemite."[28]

While all national parks are required to have a general management plan (GMP) outlining their intentions, none have inspired as much interest or have been cited as much as the 1980 Yosemite GMP. In honing its plan, the agency spent years informing and consulting the public. Deliberations included eighty-four public hearings, an extremely detailed working plan, and 62,000 participants.[29] During deliberations, the agency made public its own preference for changing the policies regarding automobiles in the valley, and, in a sign of things to come, the media began to frame the proposal in terms that its opponents would exploit from that point on. In the summer of 1980, in a front-page story, the *New York Times* reported that NPS "officials are wondering if they may soon have to restrict access."[30] However, the article did not provide the rest of the context: that the NPS would not really restrict access but rather make people make choices regarding how they would access the park—for example, by bus instead of car. Newswire organizations reported the plan in similar terms. For instance, in late October, UPI headlined a story that ran nationwide "Park Service Planning Ban

Table 3-2. *Recommendations of the 1980 Yosemite General Management Plan*

Recommendation	*Current*	*Proposed*
Overnight visitors (developed areas)	n.a.	15,713
Overnight visitors (valley)	n.a.	7,711
Day visitors (valley)	n.a.	10,530
Accommodations (valley)	1,528	1,260
Day parking spaces (valley)	2,513	1,271
Campsites (valley)	872	756
Summer employee beds	1,510	480

Source: U.S. National Park Service, *Yosemite National Park General Management Plan* (Washington: Department of the Interior, 1980), pp. 17, 25, 35.

on Cars in Yosemite Valley."[31] The story did not discuss the provision of alternative means of getting into the park (for example, regional buses); instead it stressed "significant immediate reductions of private vehicle traffic and an eventual ban on all such traffic."

When finally approved in 1980, the GMP contained some strong, explicit goals, regarding vehicles in particular. In the second paragraph, the plan decries "a march of man-made development in the Valley," then states that "the intent of the National Park Service is to remove all automobiles from Yosemite Valley and Mariposa Grove and to redirect development to the periphery of the park and beyond."[32] The plan contained recommendations to pursue more than 300 specific actions affecting a wide range of issues in the park over the next ten years.[33] Regarding transportation, the GMP restated the emphasis on eliminating automobiles: "The ultimate goal of the National Park Service is to remove all private vehicles from Yosemite Valley. The Valley must be freed from the noise, the smell, the glare and the environmental degradation caused by thousands of vehicles."[34] Importantly, though, the NPS also planned to provide alternative ways to get people into the valley.

Table 3-2 shows the recommendations of the 1980 GMP. To get cars out of the valley, the agency would need to set up parking areas in regional locations and then bus visitors to their destinations.[35] That would take time. Meanwhile, the NPS would begin to implement reductions in traffic by enforcing established capacities regarding accommodations and parking. As table 3-2 shows, the agency proposed substantial cuts in parking spaces, accommodations, and housing for employees. For a time frame, the agency used "the next 10 years" as a framework but stated "the immediate plan is to greatly reduce traffic" in the valley.[36]

Paving and Good Intentions

For all its intentions, however, the NPS was not able to stop paving the way, literally and figuratively speaking, for automobiles to access the park. While agency personnel had finally clarified their own intentions for the park, they had not attracted strong support from external policymakers. Critics claimed that reducing automobile use would restrict freedom of access, and the agency itself did not provide extensive economic analysis or scientific studies supporting the proposed recommendations.

Perhaps the most important evidence of lack of external support for the plan was lack of funding. A 1986 project study estimated the cost to implement the 1980 plan as at least $70 million, including $22 million for new parking areas outside the park, $33 million for new buses, and $19 million for annual operations. The team, which included officials from the park and from regional headquarters, admitted that "in the foreseeable future, however, it is doubtful implementation could occur due to high costs."[37] The annual total budget for the park in the 1980s and early 1990s was only a fraction of the amount needed to make substantial changes. In an interview with me in 1992, Yosemite superintendent Mike Finley blamed the slow pace of implementation of the transportation plan almost entirely on low financial resources: "Our annual budget is only $15 million here, and that project ran over $70 [million]."[38]

In addition to providing little financial support, Washington policymakers showed little interest in pursuing the 1980 proposals, and many showed more interest in protecting the activities of private concessions operations in the parks. In 1990 hearings of the Senate Subcommittee on Public Lands, for example, Secretary of the Interior Manuel Lujan Jr. questioned the low fees charged to concessionaires, particularly those operating in Yosemite. His comments drew the wrath of committee members. Senator Malcolm Wallop (R-Wyoming) stated that "Secretary Lujan's appraisal of the situation is a completely ignorant view of ordinary business practice in America."[39] Nor did the Ronald Reagan or the George H. W. Bush administrations show much interest in pursuing the ambitious goals of the 1980 GMP. If there was any momentum under the plan, it came to a halt under the tenure of James Watt as secretary of the interior. Indeed, Watt stated explicitly that concessionaires would have a greater say in park decisionmaking than they had in the past and even cited Yosemite as an example.[40] As NPS chief of cultural resources Jerry Mitchell told me, "the Reagan administration's priorities were for facilities, and that kept the agency's priorities from working up the list."[41]

Some high-ranking NPS personnel seemed resigned to the status quo. In 1984, for instance, Yosemite superintendent Robert Binnewies stated that "a certain level of development has been achieved within the park and that level should now be held."[42] In 1987, the new superintendent, John Morehead, characterized further efforts to move facilities outside of the valley as impractical.[43] In 1990, NPS director James Ridenour referred to the 1980 plan as only "a concept, a good ideal,"[44] a view that was consistent with that of his political bosses.

Given the lack of strong political support, NPS personnel settled for pursuing incremental progress on achievable projects. The 1986–87 study team concluded that it made more sense that "the limited personal and fiscal resources of the NPS will henceforth be put into projects that have a high probability of being accomplished and that are, indeed, steps in the implementation of the GMP."[45] Regarding transportation, the study team concluded that "currently, it is not considered practical to remove all 1,000 parking spaces called for in the GMP because that would result in a shortage of spaces to the daily visitor use levels described in the GMP."[46] NPS strategic planner Chip Jenkins came to virtually the same conclusion: "Over a period of twenty years, you can make significant impacts on the park through small, incremental changes. The priorities of the country won't allow us $300 million at once; rather you need to do a whole slew of smaller projects that added together will make the park better."[47] Agency personnel replaced some asphalt and concrete in the village with trees and shrubs. They extended

Table 3-3. *Development of Yosemite Valley in the 1980s*

New concessionaires since 1980	*Expanded concessionaires since 1980*
Camp Curry pizza and ice cream	Marketing and catering conference services
Raft rental	Bicycle rentals
Wine tasting and music festivals	Cross-country ski rentals
Photo-finishing center	Yosemite Lodge
Video rental	Bracebridge Dinner
Tobacco store	Village Store
Camp store at ice rink	

Source: Senate Subcommittee on Public Lands, National Parks, and Forests, *Concessions Policy of the National Park Service,* 101st Cong., 2nd sess. (Government Printing Office, 1990), p. 110.

bikeways and removed some power lines. They built and installed more bear-proof lockers and dumpsters.[48]

Unfortunately for change proponents, even as the NPS was taking those small steps during the 1980s, the park witnessed more unwanted growth. Table 3-3 uses data from the NPS and the Wilderness Society provided in congressional hearings to show the expansion of visitor services in the years following the 1980 GMP. The changes mentioned in the table are not enormous, but they do constitute greater use of the park's resources. Excessive rafting on the Merced, for example, can result in litter problems in the river and erosion of its banks.[49]

Into the 1990s, policymakers at Yosemite had taken only small steps toward the goals proposed in the 1980 GMP. Some argued with some pride, however, that they had for the most part halted further expansion of commercial services and even more development in the park. In 1996, two different officials both told me proudly that the period between the 1980 GMP and 1996 had seen the least amount of change to Yosemite of any period in the park's history.[50] Other observers countered that holding the line on development was not the same as returning the park to a more natural state. As historian Alfred Runte concluded in his 1990 book, it seemed that "Yosemite National Park in the future would look much like it had looked in the past."[51] When I asked Runte in 1992 if he anticipated much success in restoring natural conditions in the park, he answered simply, "No, business as usual."[52] The NPS had made and was pursuing some changes, albeit mostly incremental adjustments, to the natural conditions in the valley, but, particularly with regard to automobile traffic, the agency had definitely backed off its 1980 promises.

A Plan to Free the Valley

The irony was not lost on me. I was writing about the impact of cars on the Yosemite Valley, and yet here we were driving a car into the valley on one of those summer days when you expect to see congestion and long lines of traffic. My wife and I had camped with my brother and a friend and his two sons in the relatively quiet Tuolumne Meadows the night before. But that July morning we did not see the lines of traffic that we feared; instead, we drove straight to our campsite and parked the car, which was not to be used for the next couple of days. Like others who now camp in the park, we had gone through the ordeal of getting a campsite, phoning in as soon it was possible to reserve one, months before. After two hours of trying, we had failed to get a spot for June, but we got one of the precious sites for July. As we looked around at our gorgeous surroundings, we could imagine no reason why we would want to use the car again until we left.

A Catalyst for Change

The fact that campgrounds are so scarce in Yosemite is a reflection of a larger story. In 1996, when I met with Hank Snyder, the NPS chief of resources for the park, we were talking about how difficult it is to change behavior substantially in a place like Yosemite. He pondered the issue for a moment and then offered, "You need a catalyst to do that."[53] As mentioned in the first chapter, major policy changes are much more likely to be made when an existing policy system is shaken by some unplanned external event, for example, a natural disaster.[54] By triggering different ways of thinking or behaving, such unplanned jolts can serve as a catalyst for change. Whether such an event occurred in Yosemite with the 1997 floods remains to be seen. At the least, the current number of campsites (464) is about half the number that existed before the floods. And the floods did increase the momentum for change, which resulted in the 2000 Yosemite Valley Plan. That plan renewed the "ultimate goal of freeing the Valley of the environmental and experiential degradation caused by thousands of vehicles."[55]

Even before the 1997 floods, people involved with Yosemite were expressing frustration with the slow pace of implementation of the 1980 GMP. Those frustrations inevitably focused on transportation, for reasons obvious to anyone who entered the park in the 1990s. As Chip Jenkins told me in 1996, "the major problem with the park is the movement of people, and most movement is by automobile."[56] On an average summer day in the early 1990s, as many as 20,000 visitors entered the park in approximately 7,000 vehicles;[57] on Saturdays, the number jumped to nearly 7,400 cars.[58] Traffic problems occurred regularly on roads leading into the park and in places like the Mariposa Grove, but they were most severe in the valley. The agency estimates that at least 70 percent of all summer visitors to Yosemite travel to the valley,[59] which alone averaged more 10,500 day-use visitors and 7,700 overnight visitors each day in the early 1990s.[60] Only 15 percent of those visitors came by bus. [61] While the 1980 GMP called for reducing the number of parking spaces in the valley by nearly half—to 1,271—the number had actually increased to more than 5,000. On busy weekends, the NPS occasionally had to close the gates, allowing one car in only after one left.[62] The lines of traffic were not only obnoxious to those in or near the cars, but they contributed significantly to air and noise pollution in the valley.

Federal policymakers showed some interest in changing the situation. In the Intermodal Surface Transportation Efficiency Act of 1991, Congress called for a study of alternative methods of transportation in the national parks. Planners used Denali, Yellowstone, and Yosemite for the case studies. The Yosemite study did not recommend an alternative, but it did provide an assessment of options. One alternative was to use a regional approach, having visitors park their vehicles in outlying communities such as El Portal and then ride buses or light rail into the park. That option would require regional cooperation and a high level of financial resources. Planners estimated annual operating and maintenance costs at $3 to $5 million for a bus system and $5 to $6 million for rail transit. Capital costs for constructing the necessary facilities would run about $26 million for buses and $279 to $459 million for rail.[63] Another alternative was to let people drive into Yosemite and

In 1997 floods reached record levels in the valley, as indicated by the sign in this meadow in the foreground of Yosemite Falls. The floods were a catalyst for rethinking policies on the park.

then park at staging areas to access the valley. Costs to construct and operate such facilities varied by location, but the least expensive option would be to build a 1,800-vehicle, $11 million parking structure in the west end of the valley at a place called Taft Toe; visitors would be required to ride shuttles from there to their destination.[64] Another possibility was to restrict the number of visitors to the park. The National Park Service had never before attempted such an approach on such a scale and at so popular a place. Agency officials hesitated to take that step, not wanting to deny people a chance to visit.

While those options were being discussed, none received any kind of official endorsement. Some were ridiculed. The possibility of building a new parking structure in the valley, for instance, stirred strong opposition from environmental groups as well as the local congressman, George Radanovich (R-California), who termed the idea "an assault on common sense."[65] The NPS therefore did not commit itself to making any major changes in transportation immediately; instead, it began to participate in the Yosemite Area Regional Transportation Strategy (YARTS) process and continued to consider various options.

Instead of formulating a whole new plan for change, agency personnel continued to build on the modest steps that they had taken in the 1980s. In the early 1990s, the NPS imposed occasional bans on daytime campfires in the valley to try to improve the air quality, which was suffering from the number of fires and cars. The agency moved some facilities to El Portal to reduce the need for some employees to drive into the valley.[66] In 1995, the agency added two electric buses to the shuttle system to replace the polluting diesel buses. The agency continued to emphasize the use of the shuttle system, which had been moving people throughout the valley, on a voluntary basis, since the 1980 plan was presented. On an average summer day in the early 1990s, the shuttles would carry more than 22,000 riders (that figure includes some individuals who rode the shuttles more than once).[67]

Then, at the start of 1997, the floods came. Three days of heavy rain, on top of heavy snowpack that already was melting from relatively warm weather, produced the biggest floods in the park in eighty years. The Merced River, flowing through the center of the valley, rose eight feet above flood stage. The impacts were dramatic. The event closed the park for several months, flooded hundreds of acres of meadows, submerged half of the valley's 900 campsites under water and silt, and destroyed a huge amount of infrastructure, from bridges to picnic tables. Some estimates put total damages at $178 million.[68] Evidence of the flood can still be seen in signs that depict the high-water mark in places that no one ever expected to see under water.

The flooding also provided an opportunity. In describing that opportunity, pro-change advocates used the same term to describe the event that Hank Snyder had suggested to me the year before: "A 1997 flood provided an unexpected *catalyst* to revamp NPS' approach to development at Yosemite [emphasis added]."[69]

Some of the immediate responses to the floods hinted at what was to come. First, in February, the NPS announced that for the first time in the history of the park, it would require reservations for people to enter in the summer. Simply put, the flood had damaged enough of the infrastructure that the park could not handle the usual hordes of visitors. The agency set up a toll-free telephone number for people to use to make reservations. Second, the agency began considering rebuilding and widening the El Portal road (a major artery into the valley) to make it safer for bus travel. Third, Congress anted up $176 million to help recovery. The funds could be used to rebuild, but they also could be used to carry out goals from the 1980 plan that were now suddenly more attainable, even if it meant leaving areas natural that had previously been developed. One key example involved an area, roughly 200 acres, between Camp Curry and Yosemite Village. The flood washed away the campground in that spot, and NPS personnel recognized the opportunity to not rebuild it. A fourth impact was more complicated. The agency renewed its efforts to develop a plan for the valley, taking advantage of the head start that Mother Nature had just given it to recreate natural conditions in the valley. But the plans for change soon ran into detours.

The agency split up the planning for the valley. One area of the park that had been severely damaged by the flood was Yosemite Lodge, which provides housing for visitors as well as for many employees. The NPS separated out the lodge redevelopment plan from the rest of the plans and put the former on a faster track. Specifically, the agency called for building a new structure outside of the floodplain, in the area known as Camp Four (or Sunnyside). Camp Four is important to a lot of people, particularly climbers. Yellowstone is world-renowned for climbing, and it is a magnet for climbers of all skill levels.[70] Camp Four is therefore a colorful, diverse, and international place. It's also the only walk-in campground in the valley and often serves as a last resort (pun intended) for campers who did not plan far ahead or bring a fat wallet. The NPS also jumped ahead of the ongoing YARTS transportation planning process and proposed building the new parking structure at Taft Toe. Further, agency personnel continued developing a separate plan for the Merced River. In 1987, Congress had designated 122 miles of the Merced, eighty-one of them within Yosemite, as a Wild and Scenic River; such a designation carries with it the mandate for a management plan.

The separation of plans and the content of individual plans worried various constituencies and led to legal actions. Climbing groups, the Friends of Yosemite Valley, and the Sierra Club all sued. The plaintiffs included Yvon Chouinard, a famous climber and the founder of the Patagonia sports clothing company. In October 1998, U.S. district court judge Charles Breyer issued a temporary injunction against the lodge redevelopment plan.[71]

The agency did, however, continue to develop the Merced River Plan, independently, for the most part, of the subsequent valley plan. The agency conducted some planning workshops in the mid-1990s to begin to develop management strategies for the river corridor, but that too was affected by the courts. Specifically, the plan to rebuild the El Portal road would affect the Merced River corridor, and the district court ruled that "the absence of a river management plan hindered the National Park Service's ability to ensure that projects in the river corridor adequately protect the Merced Wild and Scenic River."[72] The NPS obviously needed to develop an overall plan for the valley.

A Plan for Change

The ultimate product of all the discussions in the 1990s and the catalyst provided by the 1997 flood was the 2000 Yosemite Valley Plan (YVP). The YVP did not replace the 1980 GMP in that the GMP addressed the entire park while the YVP focused on the valley. Nor did the YVP replace the Merced River Plan, released in August 2000 to guide management of the river corridor. It did not replace several other plans either, such as those relating to concessions, but the NPS did intend for the YVP to be the coordinating and guiding document for Yosemite Valley. As such, the YVP reiterates, at least rhetorically, many of the principles guiding park management, which date back to Olmstead but were stated more recently in the 1980 GMP. Once again, the issue of cars in the valley was prominent. The most specific of the five broad goals in the YVP (in addition to reclaiming natural beauty, allowing natural processes, promoting visitor opportunities, and reducing crowding) is to "markedly reduce traffic congestion."[73]

The agency formulated the Yosemite Valley Plan through the environmental impact statement (EIS) process required by the National Environmental Policy Act. The process involved scoping, development of alternatives, and public comment. Scoping took place throughout the late 1990s with public identification of issues and alternative approaches. The agency then produced a draft Yosemite Valley Plan in April 2000. The draft plan provided five alternatives (including "no action") and public comments were solicited.[74] The managers of Yosemite engaged in an intensive effort to involve the public in their deliberations; in the April 2000 planning update, for instance, superintendent Dave Mihalic urged any interested parties "to feel a sense of ownership and take part in protecting the Valley."[75] During the comment period, which took place between April and July of 2000, the agency received more than 10,000 written comments and held fourteen public meetings in California alone, drawing nearly 1,500 people.[76] The agency also held hearings in Seattle, Denver, Chicago, and Washington, D.C.

Besides the no-action option, the agency offered four comprehensive alternative policies for managing Yosemite Valley, which are summarized in table 3-4. Several factors should be kept in mind. The figures shown in the no-action column reflect conditions after the 1997 flood; the figures shown in the other columns are what would exist if the alternative was implemented. Keep in mind that the valley contains about 3,500 acres. All four alternatives to taking no action would relocate headquarters for the NPS and concessionaires outside of the valley, relocate the stables to Foresta, expand shuttle bus routes, and make

Table 3-4. *Alternatives in the Yosemite Valley Plan*

Alternatives	1 *(No action)*	2 *(Preferred)*	3	4	5
Acres restored to natural state	0	176	209	194	157
Parking for day visitors	1,600 (scattered)	550 (village); 1,570 (out of valley)	1,622 (Taft Toe); 0 (out of valley)	550 (Taft Toe); 1,600 (out of valley)	636 (village); 1,080 (out of valley)
Overnight parking	1,929	1,721	1,659	1,627	1,801
Total campsites	475	500	449	441	585
Total lodge units	1,260	961	982	982	1,012
Employee beds (valley)	1,287	683	689	689	752
Ahwahnee/ Stoneman Meadows	Keep roads	Remove roads	Remove roads	Remove roads	Keep roads
Northside Drive (Lodge to El Capitan Bridge)	Keep road	Close road and convert to trail	Close road and convert to trail	Close road and convert to trail	Convert one lane to trail
Developed acres in valley	407	336	335	341	343
Visitor center	Retain in current spot	New facility near parking	New facility at Taft Toe	New facility at Taft Toe	Retain in current spot
Total cost of changes	0	$457 million	$421 million	$454 million	$495 million
Transportation change cost	0	$83 million	$38 million	$81 million	$79 million

Sources: U.S. National Park Service, *Draft Yosemite Valley Plan: Supplemental Environmental Impact Statement* (Washington: Department of the Interior, 2000); U.S. National Park Service, *Yosemite Valley Plan: Final Supplemental Environmental Impact Statement* (Washington: Department of the Interior, 2000).

some minor changes, such as removing the Ahwahnee tennis courts. The four alternatives, besides no action, are roughly comparable in terms of total costs.[77]

The important differences between the options involve transportation. Alternative 2, the preferred alternative, involves the highest transportation and circulation costs. It would not construct a new parking facility at Taft Toe, it retains some in-valley parking, and it moves some parking out of the valley (420 spaces at Badger Pass, 780 spaces at South Landing, and 370 spaces at El Portal). Alternatives 3 and 4 would require a new parking facility at Taft Toe. Alternative 3 would not require out-of-valley facilities; its transportation costs therefore would be less than those for alternatives 2 or 4. The preferred alternative differs from alternative 5 in removing more roads and facilities while retaining valley parking. The NPS anticipated that the preferred alternative, when implemented, would eventually reduce the number of vehicles driving in the east valley (the Half Dome end) by half. The plan also called for expanded shuttle service. Shuttles from the new visitor center to the Ahwahnee would run every fifteen minutes; to the west valley (Bridalveil Falls), every seven and a half minutes; and to the east valley (Curry Village and campgrounds), every four minutes. Out-of-valley shuttles would run from Badger Pass and El Portal every twelve minutes and from Foresta, every six minutes.[78]

In the YVP, the NPS admitted that it had "considered but dismissed" several other alternative strategies. While the agency's candor was admirable, others would later question its reasons for dismissal. Indeed, the dismissed alternatives included options that would have been the choice of some advocates for aggressive change, perhaps the most obvious being the option to simply mandate removal of all private vehicles from the valley. The NPS admitted that total removal was, after all, "the ultimate goal of the 1980 GMP" but nevertheless dismissed it as "economically infeasible and impractical at this time," adding that "a phased, collaborative approach would be required to achieve this goal."

One way to move toward such a goal would be to provide no parking for day visitors in the valley. The NPS also dismissed that alternative as "economically infeasible," given the cost of the fleet of transit vehicles that would be required to get visitors into the valley.[79] Another obvious alternative would limit the number of visitors to the park by requiring reservations. That one was favored by environmental activists like David Brower, who argued that "placing no limit on the number of current visitors who can visit the park at one time is a violation of the Organic Act and a breach of our contract with future generations."[80] The NPS dismissed that alternative with somewhat vague reasoning, stating that "it does not solve other resource and infrastructure needs associated with this planning effort." Another alternative would return the campgrounds to pre-flood conditions rather than restore a more natural environment; the NPS dismissed that option on the grounds that it would be inconsistent with the "guidance provided" in the Merced River Plan.[81]

Although, as mentioned before, the agency tried to separate the Merced River Plan and the Yosemite Valley Plan as much as possible, over time the overlap between the two became increasingly evident. The Merced River corridor is, after all, home to many of the roadways and much of the development in the valley. In a process similar to that for the YVP and undertaken almost simultaneously, the NPS conducted scoping analyses for the Merced River in 1999, held six public meetings, offered five alternative plans in a draft EIS in January 2000, and solicited comments.[82] The agency incorporated the comments, released the plan in June 2000, revised it again, and then released what it thought was the final version in February 2001. That version states explicitly that it "does not specify detailed actions, but provides broad guidance."[83] The plan has been the subject of intense litigation, discussed later in this chapter.

The YVP became official before the end of 2000. Before recommending its preferred alternative, the agency made some changes to the draft in response to public comments; for example, it increased the number of campsites and reduced the number of higher-cost lodge units. The agency also adjusted its plan for out-of-valley parking on the Big Oak Flat road.[84] After the changes, regional director John J. Reynolds of the NPS Pacific West region signed the record of decision for the YVP on December 29, 2000. Notably, the signing came near the end of the Clinton administration and shortly before the Bush administration took office, and some critics suggested that the NPS hurried the plan to get it approved while a potentially more sympathetic administration was still in office.

Whether or not that was the intent, ultimately agency personnel did not create conditions conducive to obtaining substantial public support for their preferred changes. Pro-change advocates did not build an effective coalition through positive issue definition, convincing economic or scientific arguments, or smooth interjurisdictional cooperation. The failure to create such conditions did not lie entirely with the NPS but also with those who wanted to block the proposed changes and those who preferred different changes.

The Lockout Image

The confusion over the definition of possible restoration activities at Yosemite derives from the inherent tension in the mission of the NPS, which the agency has faced the since its creation. As noted in chapter 1, the agency is supposed to restore and preserve natural conditions in parks for perpetuity even while enhancing visitor access and enjoyment. While the NPS and its supporters attempted to frame the YVP as enhancing the experiences of both present and future visitors, opponents framed the plan as an attempt to restrict visitor access. They were able to do so in large part because of the importance of transportation to park users. As the NPS admitted in the YVP, transportation is "one of the most fiercely debated and frequently commented on subjects [because] for these individuals, true enjoyment of the park entails entering and moving about the park in certain ways."[85] Even though agency personnel argued that a change in transportation modes would simply mean riding a bus instead of driving a car to access what would then be a more natural setting, opponents claimed that it restricted their freedom. And indeed, for many visitors to national parks, cars have long been the way to see them efficiently. As historian David Louter noted, "cars have been central to our understanding of national parks as wild places."[86]

The roots of the difficulties in promoting a positive image of possible changes in Yosemite were thus deep and numerous. As mentioned, media coverage of the 1980 GMP often reported the plan as an attempt to ban cars, with little or no discussion of the plan to provide alternative means of transportation. In addition, lawsuits in the early and mid-2000s over the separation of the different valley plans and the lodge redevelopment plan itself had created confusion regarding the agency's overall intentions.[87] Furthermore, the agency had not clarified some important elements of its own plan as it evolved, in part because they were to be worked out in the future. For instance, the role of YARTS had yet to be defined.[88] One National Parks Conservation Association spokesperson sympathetic to the NPS said as early as 1998, "I think a lot of people's concerns about the plan are based simply on fear of the unknown."[89]

As publicity surrounding the plan increased, opponents of alternative transportation systems and some members of the media increasingly framed the possible policy changes as a matter of restricting access. For instance, the director of the chamber of commerce of eastern Madera County, a gateway community, voiced his suspicion that planned reductions in the number of vehicles entering the valley were really a smokescreen for the NPS's desire to reduce the number of visitors.[90] Media coverage of the evolving plan in early 2000 played into the hands of such opponents. In March, the Associated Press released a story with the headline "Yosemite Park Plan Calls for Cutting Traffic by 60 Percent."[91] The 60 percent figure was based on a comment by Interior Secretary Bruce Babbitt, with little discussion of contextual issues such as the fact that people would be able to access the park by other means. Another story, in the *New York Times,* referred to the "oxymoronic task for the National Park Service to keep Yosemite fully accessible yet return it to its pristine natural beauty," thereby implying that pursuing the latter would inevitably reduce access when in fact the agency wanted to keep the park accessible but in ways that did not rely on private automobiles.[92]

Even after completion of the EIS, critics of the NPS's efforts to change policies have continued to frame the issue as one of limiting access. During congressional hearings in early 2001, for example, the president of the park concessionaire, Delaware North Park Services Inc., warned of a "perception that Yosemite National Park is not open and accessible to private vehicles."[93] Representative George Radanovich (R-California), an important player in Congress, offered frequent warnings against "locking people out of the park."[94]

Economic Uncertainty

Agency planners faced a challenging task in addressing the economics of possible changes in valley transportation options. As stated in the

YVP, "many members of the public ask the PS to carefully consider the effects of proposals on social and economic environments, especially those of gateway communities. Many believe these towns have invested their future economic well-being in meeting visitors' needs."[95] Data on neighboring communities reinforce such beliefs. In Mariposa County, for example, just west of the park, the lodging, food and beverage, and service industries accounted for nearly 50 percent of employment in 1998, as they did in Mono County, on the east side of the park.[96] Those sectors also provided almost 30 percent of employment for the larger region as a whole.[97] Given that unemployment rates in the region (13.1 percent in 1998) were well above state and national averages, the lodging, food and beverage, and service industries were vital to the area economy. Tourist spending was crucial. For instance, transient occupancy taxes in 1998 (such as from hotels) contributed 57 percent of the revenue in the Mariposa County general fund and sales taxes brought in another 12 percent. Tourist spending per capita in the region overall varied in 1998 between $25.54 a day for day visitors and $66.68 for local overnighters.[98]

To assess the impact of possible changes on the regional economy, agency planners used current statistics on the region's economy and then projected changes in visitation totals and visitor spending, based on surveys, for all the alternatives in the EIS. They then used multipliers to anticipate impacts on regional revenue and spending.[99] The NPS obviously faced a difficult task in trying to assess future behavior. As officials observed, "projecting the magnitude and nature of future day visitation is difficult due to the complexities associated with the proposed alternatives and numerous uncertainties associated with other independent factors that may affect future visitor demand for park access."[100] After balancing various factors that might increase or decrease future visitation, the agency based its calculations on the fundamental assumption that visitor demand would "remain unchanged in the future from its current conditions;"[101] planners argued further that visitation patterns and behavior would be unchanged.[102]

Even while being sympathetic to the complexities involved in such an exercise, one can easily question specific aspects of the agency's calculations. For instance, under "factors potentially increasing visitation," the planners list increased population growth and increased tourism. Listing those factors assumes that they have contributed to increased visitation in the past, and that certainly was true in the twentieth century. But, in fact, as figure 1-1 shows, visitation to Yosemite has fallen since peaking at more than 4 million in 1996. In the category of factors possibly decreasing visitation, the planners acknowledge actual changes—not just projections based on assumptions about population growth—that would result from implementation of the YVP, such as "decreased in-park accommodations." Another calculation recalls the framing issue described above. The NPS balances "unfavorable publicity" with "favorable publicity" from marketing efforts, but can the agency claim to have been so effective in recent years that the calculation is a wash?

If even sympathetic observers can be skeptical of these analyses, imagine the general reaction. To put it mildly, local businesses and communities were not convinced by the economic analyses in the YVP environmental impact statement. The agency brought on some of the skepticism itself by its use of some confusing language. On one page of the YVP, for instance, it says "no change" to visitation patterns but two pages later, it admits that "following implementation of each alternative, visitation patterns to the park will likely change."[103] Planners did make the point, and probably justifiably so, that decreasing the number of overnight spots in the park might well increase the overnight lodging opportunities for neighboring communities, but the local business interests were not convinced. In the comments section of the EIS, the agency noted that it had received considerable criticism. Respondents expressed concern about the possibility of decreased tourism in general, lost drive-through traffic, and decreased lodging; confusion over the methodology used in the socioeconomic analysis; and concern about the admitted lack of a benefit-cost analysis.[104]

Insufficient Scientific Information

The larger context for the proposed changes at Yosemite involves alternative forms of transportation. Obviously, the debate about using public transportation—buses, shuttle systems, and light rail instead of cars—goes far beyond the boundaries of this national park. Scientists and managers could point to other parks where such systems had been used effectively. Visitors accessing the interior of Denali National Park, for example, have relied on buses for decades. Perhaps the example most relevant to Yosemite was at Zion National Park, where in 2000 the NPS began using a shuttle system that includes twenty-one buses to take visitors from a parking area on the border of the park to the valley and the popular tourist sites. This system effectively replaces the traditional pattern of transportation, which involved roughly 5,000 cars accessing the park each day and searching out (or waiting out) one of only 500 available parking places. But scientists and planners at Yosemite still had to make the case that their proposed changes would result in substantial benefits for this specific park.

One of the primary arguments for reducing the number and use of vehicles in Yosemite Valley was that doing so would improve air quality; indeed, the agency considered it "imperative that plans for Yosemite Valley stress air quality protection."[105] Air quality, then, was a major focus of the scientific analysis in the YVP. The NPS conducted air quality impact analyses for each alternative scenario in the YVP, assessing emissions from vehicles operating in the park (based on projections), and from construction and demolition activities. California Air Resources Board computer models were used to quantify emissions; air quality monitoring stations also were used in and near the park to assess levels of criteria pollutants, notably the particulates and ozone that can result from vehicle use.[106]

In short, the NPS argued in the EIS that changing transportation patterns would improve air quality in the park. The data showed that air pollution levels in the valley often were substantial and that under current conditions—and thus under the no-action alternative—the ambient air quality standard for ozone was occasionally exceeded (between three and eleven days a year) as was the standard for particulates (between one and five days a year).[107] Adopting one of the proposed alternatives, on the other hand, would reduce automobile traffic to the point of improving air quality. Anticipated reductions in pollutants under the various proposed alternative actions ranged from 10 percent to 83 percent.[108]

As with the economic analysis, the conclusions in the scientific analysis were defensible but not definitive. The agency admitted that "although mass emissions are provided for comparative purposes, the impact of an individual alternative on the ambient air quality standard in the region was not quantified [because] air quality is a regional issue that is influenced by factors outside the immediate area."[109] In other words, the issue is so complex that future projections are tenuous. Not surprisingly then, many comments in the EIS expressed frustration with the lack of conclusive data. Critics also questioned the lack of in-depth analysis of the impact of diesel fuels used in buses and the complex language used in discussing air quality in the document.[110] Change advocates could dismiss such objections as predictable, and the agency could contend that its analysis was as thorough as possible, but one comment was especially revealing. On July 12, 2000, the U.S. Environmental Protection Agency questioned the scientific analysis of the impacts from the preferred alternative on the basis of insufficient information.[111] The federal office in charge of regulating air pollution in the United States had declined to endorse the science regarding air quality in the YVP, a fact that did not bode well for attracting wide public support for the restoration plan.

Socioeconomic Impact on Gateway Communities

Those seeking changes at Yosemite did not seem, at first glance, to need as much cooperation from other institutional partners as those at the Everglades or the Grand Canyon. Many of the changes would occur within the park, largely within the valley itself; the temptation therefore existed not to fully engage other entities in the planning process. However, the proposed changes did affect areas other than the park that were in different political jurisdictions. The comments section of the YVP provided hints that interjurisdictional tensions could be problematic.[112]

Most consequential of the other entities were the planning boards or supervisors of gateway communities. Yosemite is bordered by Madera County to the south, Mariposa County to the southwest, Tuolumne County to the north, and Mono County to the east. Population centers in those areas would be affected by altered traffic plans if the changes went through; some would be affected directly by expanded parking facilities and transportation hubs. Many area officials expressed anger or disappointment at what they perceived as their lack of input in or impact on the planning process. A comment from the Mariposa County board of supervisors is illustrative: "the conclusions contained in the Valley Plan relative to the socioeconomic impacts of relocation are minimized due to the inadequate understanding of the nature of the communities and how services are delivered."[113]

Some evidence also suggests that the NPS was in a lonely position with respect to proposed changes, even among its federal counterparts. In the comment section of the YVP, the supervisors of the neighboring Sierra and Stanislaus National Forests also criticized the recommended alternative in the YVP because of likely consequences (for example, increased use of national forest lands) "that have not been adequately mitigated."[114] In other words, pro-change advocates did not constitute a coalition for change that included agency officials from potential institutional partners in the proposed actions.

Getting It Right, a Green-Wash, or Nonscientific?

Conflicts in framing the issue, questionable science, contested economics, and interjurisdictional tensions associated with the proposed changes in the YVP precluded stronger and broader public support of possible repairs. That is not to say that no participants or observers were enthusiastic about the plan, but the NPS did not enjoy nearly the support that it needed.

Many people did applaud the potential changes to Yosemite Valley. At the time, a survey by an independent firm showed that 81 percent of those polled favored removal of unnecessary development in Yosemite Valley.[115] The NPS cited public opinion surveys that showed that more than 80 percent of park visitors supported day use visitor parking and 76 percent responded positively to the question "Would you be willing to park your car outside the park, one-half hour or more away, and take a shuttle bus to and from Yosemite?"[116] A statewide poll showed strong support for both the YVP and the Merced River Plan.[117] Several major environmental groups endorsed the YVP. For example, Jay Watson, a regional director of the Wilderness Society, said that "the Park Service should be applauded for listening to the public."[118] Watson later defended the plan as "the product of one of the most honest, open, and rigorous planning processes ever undertaken by the Park Service."[119] Janet Cobb, the president of the Yosemite Restoration Trust, commended the NPS and added, "We are looking forward to implementing the Yosemite Valley Plan."[120] Several major news organizations, including the *New York Times,* also applauded the plan,[121] and at least one observer perceived "a consensus that the Park Service had finally gotten it right."[122]

That perceived consensus was illusory. Others felt that the proposed changes did not go far enough. Environmentalist David Brower called the draft a "green-wash for a half-baked development plan."[123] Others criticized the fact that whereas the 1980 GMP prescribed maximum daily use totals, the 2000 YVP "does not propose specific limits on visitation."[124] Instead, the agency promised a Visitor Experience and Resource Protection (VERP) study within five years to develop recommendations for the level of visitor use of each area of the valley. Further, the plan stated plainly that, even under the preferred alternative, "out-of-Valley shuttle buses would not be ordered for years."[125] In addition, critics bemoaned the fact that in contrast to the 1980 GMP, the plan was now to "reduce traffic" but not eliminate cars from the valley. Shortly after completion of the YVP, the executive director of the Friends of Yosemite Valley said that it "holds no promise of restoration."[126] Even some proponents of the plan admitted that it was only "a modest first step."[127]

Not surprisingly, others criticized the plan for going too far. Neighboring communities expressed the most intense opposition.[128] The Madera County board of supervisors unanimously opposed all the alternatives in the YVP. The Coarsegold Resource Conservation District within Madera County criticized the NPS recommendations as "based on incomplete or nonexistent plans, and non-scientific or professional standards."[129] The Mariposa Country board of supervisors claimed, as noted earlier, that the plan lacked complete analyses of socioeconomic impacts. The Tuolumne County board of supervisors similarly declined to support the YVP. I do not mean to suggest that all the residents or even all the businesses in surrounding communities opposed the plan. Subsequent events showed some support in these areas, but the overall reaction of the leaders of gateway communities in 2000 was anything but supportive.

To get a better sense of the mix of comments on the YVP, consider the following. As already mentioned, the NPS received more than 10,000 comments and responses during the EIS process. While agency personnel could not formally code all of them, they did code 37 percent (3,741) as being the result of organized response campaigns.[130] They coded these responses as "forms," reflecting at least some identical content. The greatest number of form responses (1,287, or about a third) consisted of comments that said "no" to the proposed parking structure at Taft Toe, an unsurprising result given that both environmental and business groups opposed the idea. The second most frequent form response (1,002 comments) welcomed the proposals to cut traffic and reduce development. The third most common response (304 comments) called for "stopping" the YVP. Since the Taft Toe issue inspired opposition from various coalitions, it can be dropped from any calculation of support. Concentrating on the latter two responses suggests a ratio of 3 to 1 in support of the YVP, a strong majority but far from unanimous.

A second means of assessing public response is to examine the origin of the comments. Of the more than 10,000 comments, at least 76 percent came from within California; I use the term "at least" since 14 percent did not identify origin.[131] Thus, less than 10 percent of comments came from outside the state. That does not constitute a base for widespread, national public support. Most of the California-based comments came from the highly populated areas of San Francisco, Los Angeles, and the Sacramento Valley. However, almost 9 percent (667 comments) came from the park and the immediately surrounding areas, including 342 from eastern Madera County alone;[132] those areas would potentially feel the greatest economic impact of changes to visitation patterns and were home to some of the most vocal opponents of the plan. It is no surprise then that the local representatives to Congress and the state legislature, the people whose support would be most essential to implementing the plan, opposed the YVP.

In fact, those key political actors expressed their opposition to the YVP strongly in 2000 and continued to do so in the following years. Two congressional districts encompass Yosemite. In 2000 the representative from the 4th District, to the north, was John Doolittle, a Republican who had been described as having "one of the most conservative voting records in the House."[133] Doolittle consistently received scores of zero from the environmental group League of Conservation Voters. Not surprisingly, Doolittle condemned the YVP; he specifically criticized the recommendation to reduce parking spaces, saying that serious traffic congestion "only exists a few days a year."[134] Doolittle also argued that the preferred alternative would have a negative impact on the economies of gateway communities. To the south of Doolittle's district, George Radanovich represented the 19th District, which contains much of the park. Radanovich, another Republican who earned a zero from the League of Conservation Voters in 1998, did express some appreciation for the agency's work on the YVP but called the planning process "confusing" and contended that it had "baffled the public."[135] Radanovich became chairman of the House National Parks Subcommittee in 2002 and thus played a crucial role throughout efforts to implement the YVP. Finally, as if the lack of federal support was not enough, California state senator Dick Monteve, who represented three of the gateway communities, also expressed concern over the YVP's possible economic effects on those communities.[136]

Political Obstacles

Shortly after arriving in the valley in July 2008, I walked through the village, past the visitor center, and up to the administration building. It was the same route I used to walk on my way to work thirty summers before. I had always loved the backdrop, the Yosemite Falls tumbling off the cliff up ahead. Indeed, I found the view inspiring even when I was reporting in just to find out which toilets I would be cleaning that day. My work on this particular day was quite a bit different, however; I was meeting with the chief of planning, Linda Dahl. I had just driven into the valley, so my opening question to her was predictable: Why, given the 1980 plan and the 2000 YVP, had it been so easy for me to do that? After all, those plans stated explicitly that "the intent of the NPS is to remove all automobiles from Yosemite Valley and Mariposa Grove."[137] Dahl acknowledged that "the 1980 GMP said get cars out of the valley" but added, "Politically, there are a lot of obstacles."[138]

The Call for an Incremental Approach

I doubt that anyone in the NPS was overconfident that implementation of the 2000 YVP would be easy; it has never been easy to implement any plan at Yosemite. As Jerry Mitchell, NPS chief of cultural resources, remarked in 1996, "history is littered with plans that went nowhere."[139] That sort of observation breeds skepticism, and even those who supported the YVP approached it warily. Immediately after approval of the YVP in 2000, Hal Browder, the vice president of the Yosemite Restoration Trust, warned, "The best plan is meaningless if it sits on a shelf."[140] Similarly, park historian Bob O'Brien endorsed the plan but warned that "full implementation will depend on people continuing to push for it."[141] Even as early as the summer of 2001, less than a year after final approval, Dave Mihalic, Yosemite's superintendent, felt the need to reassure people that the plan was alive; in that summer's planning update, he wrote that "rumors abound, but the goals and actions outlined in the Yosemite Valley Plan have been approved, not 'killed.'"[142] Mihalic and others involved with the YVP were wise to be wary. As discussed, support for the recommended alternative among the general public and political authorities was already weak. It would get weaker.

Political support for the policy changes from Congress was essential, but relevant members had already voiced their objections and continued to do so. The House Subcommittee on National Parks held a hearing in March 2001, just three months after final approval of the plan. In his opening statement, subcommittee chairman Joel Hefley (R-Colorado) set the tone by referring to "alleged" threats from automobiles and facilities and then stated, "While many people in this room would agree that the Valley may be crowded during certain peak times, many would disagree with a number of recommendations slated for action in the Valley plan."[143]

Other participants discussed the recommendations, focusing much of their criticism on the transportation component. Representatives Radanovich and Doolittle reiterated the criticisms that they had voiced in the comment period for the EIS.[144] Because members of Congress typically try to defer to the wishes of other members who are representing their home districts, their comments carried considerable weight. Officials from nearby Madera, Mariposa, and Tuolumne counties also restated their concerns about changes in the park that might inconvenience visitors and thus "adversely affect businesses" in nearby communities.[145] The critics also contended that the planning process

Yosemite Valley is host to millions of visitors each year, many of whom do not want to leave their vehicles behind, including the drivers of these RVs, lined up to obtain parking permits.

had been "fatally flawed" in that it did not allow them sufficient opportunity for input, thereby resulting in a "sham of public participation when decisions have already been made."[146]

Some contended that the Clinton administration, notably Secretary of the Interior Babbitt, had used the flood as an excuse to push its own agenda with little regard for public opinion. Radanovich, in particular, pushed NPS regional director Reynolds and Mihalic to admit that the administration had pressured them to finish the process before Clinton left office.[147] Doolittle also questioned the administration's timing.[148] Radanovich, stating that his intention was not to scrap the whole plan, recommended that "the Park Service implement the most incremental, least cost, and least disruptive elements of the flood recovery and park improvements first, and reevaluate each step as the public experiences the improvements."[149] That call for incremental behavior would prove to be the determining factor in the implementation of the YVP.

Proponents of the plan—a loose coalition but nevertheless one that did include interest groups, journalists, and agency officials—attempted to counter the criticisms. Representatives of the Wilderness Society and the Natural Resources Defense Council defended the process as legitimate and the plan as necessary.[150] Jay Watson of the Wilderness Society appended to his testimony thirty-four editorials from fourteen different newspapers in the state that supported the plan. NPS officials stood by the YVP, spending most of their time describing the reasoning behind specific aspects of the plan. Even then, however, cracks in the pro-change forces were apparent. George Whitmore, chairman of the Sierra Club's Yosemite Committee, contended that the plan did not go far enough. Whitmore declared that "the new Valley plan has abandoned the concept of limits which was in the 1980 general management plan" and called for "a reservation system for day use."[151] His demands fell on deaf ears.

Another hearing, held in 2003 on April 22, Earth Day, reinforced the skepticism of Congress and the framing of the proposed changes as restricting freedom and access. The hearing was held at Yosemite's visitor center by three members of the National Parks Subcommittee, including Representative Radanovich. Several dozen protesters of the plan showed up, some in striped prison outfits intended to symbolize their being held prisoner by NPS policies. The protesters objected in particular to the proposal to cut campsites in the valley and the related proposal to remove parking spaces. Radanovich reinforced the "lock-out" framing by declaring, "As long as I represent Yosemite National Park and this beautiful valley, I will not allow it to become an exclusive retreat available only by tour bus, nor a natural preserve which you can get to only on foot."[152] Radanovich claimed to be representing local business owners, fearful of declining visitation, but other local business owners sent a letter to Radanovich supporting the plan, claiming that it would "enhance the local economy . . . without denying access."[153] Those arguments were lost in the perceptions of "an exclusive retreat."

Pro-change forces found little support from another institutional player, the courts. Much of the focus of litigation has been the plan for the Merced River. Two local environmental groups, Friends of Yosemite Valley and Mariposans for Environmentally Responsible

Growth, challenged the Merced River Plan in court, claiming that it failed to protect the river and river corridor at a level consistent with its Wild and Scenic River status. The U.S. district court upheld much of the plan, but in October 2003 the Ninth Circuit Court of Appeals found the plan deficient for two reasons. First, the court criticized "a failure to adequately address user capacities" for the river area. Second, the plan drew improper boundaries in El Portal.[154] The NPS revised the Merced River Plan in the revised comprehensive management plan of 2005, proposing the use of the VERP framework to specify user limits and reconfigure corridor boundaries.[155] The two groups challenged that version, too, with the Friends of Yosemite Valley arguing that "VERP would allow ever-increasing use and degradation."[156] In 2006, U.S. district court judge Anthony Ishii found in their favor, ruling that the agency had again failed to prescribe actual limits on use of the corridor.

The story of the Merced River Plan is important to the YVP in several respects. First, as mentioned, the NPS had used the Merced River Plan to legitimize its dismissal of the possibility of restoring the campgrounds. Second, the actions of the Friends of Yosemite Valley indicate the passionate involvement of this group and others in any actions that affect the valley. Third, the fact that the courts have been willing to overturn the agency not just once but twice demonstrates the restrictions on the agency's autonomy in park planning. Finally, the connections between the Merced plan and the YVP cannot be ignored in discussing implementation of the latter in the coming years. Specifically, the courts have displayed a preference for seeing the agency set "maximum use capacities" and a skeptical attitude toward reliance on the VERP framework.

NPS personnel quickly realized that the Radanovich prescription for incremental changes was inevitable. NPS director Fran Mainella stated explicitly in 2003 that, at least in the following five to seven years, "there will be no net loss of day use private-vehicle parking in the valley."[157] Instead, agency employees implemented what they could, taking modest actions such as those pursued in the 1990s to facilitate natural processes in the park without making dramatic changes to traffic patterns. That has translated to slow going. Shortly after his appointment in January 2003, superintendent Mike Tollefson also was realistic about how policy change would occur in the valley. In the planning update of December 2003,

Tollefson wrote that implementation of the VYP recommendations could take twenty years and added, "We're taking it one job at a time."[158]

Reducing Impact

By the latter part of the first decade of the twenty-first century, many wondered whether the NPS would ever reduce the number of cars in Yosemite. In the introduction to a 2007 collection of essays about the park, one contributor wrote, "John Muir fought dam building, loggers, miners, and sheepherders who threatened and despoiled Yosemite in his day. We can carry on his work by reducing auto traffic and development, shielding stream banks from erosion, restoring the historic role of fires, and ultimately draining and restoring Hetch Hetchy Valley, to name just a few of the challenges we face today."[159] Notably, the first challenge he listed involved automobiles.

Some now say that the NPS had never intended to remove cars from the valley. Bob Hansen, president of the Yosemite Fund, the most supportive nonprofit organization working with the NPS, told me emphatically in 2008 that "they [the NPS] have no intention of restricting the number of cars in the valley." He further claimed that "the 1980 GMP does not say they will reduce or eliminate cars in the valley. They will reduce the impact."[160] Therefore, not surprisingly, Hansen said that his organization is "extremely happy" with the progress that the NPS has made on implementing the 1980 plan.[161] NPS veterans give a different response, one that echoes the earlier quote from chief planner Dahl about political obstacles. In 2008 one veteran interpretive ranger told me, "We'd like to reduce cars, but every time we propose something, someone takes us to court about restricting access." Another longtime employee said, "Sure, we'd like to reduce the number of cars, but that's difficult to do."

Why does implementation of the 1980 GMP and, more recently, the 2000 YVP, continue to be so problematic? Policy change proposals continue to fail to attract the support of a larger audience—and thus political authorities—to any significant degree. The NPS has yet to convince many residents of gateway communities of the economic value of making substantial changes in the park transportation system or even a lot of the public of the environmental benefits of reducing the number of cars in the valley. Further, opponents to such changes have continued to frame the issue as restriction of access and the YVP as a possible precedent for similar actions in other parks. For example, in congressional testimony in 2005, a senior analyst for the Center for Public Policy Research, a conservative think tank, argued that "if a U.S. Park Service plan to limit access to California's spectacular Yosemite National Park is allowed to stand, similar schemes could soon be in the works for other national parks, forcing many vacationers to go elsewhere for their relaxation."[162]

One obvious manifestation of the lack of wider support is lack of funding from Congress. Park superintendent Mike Finley had noted that lack of financial resources was slowing implementation of the 1980 plan. When I asked Linda Dahl in 2008 about the continued slow pace of developing regional transportation systems, she pointed to the $457 million estimate for capital costs for the preferred alternative in the 2000 YVP: "We can't even afford to clean the bathrooms in this park, and nobody is standing in line to give us $457 million."[163] Typical annual appropriations are about $24 million; in addition, under a program established in 1996 as the Recreational Fee Demonstration Program and renewed in 2004 as part of the Federal Lands Recreation Enhancement Act, Yosemite gets to keep 80 percent of the entrance fees. Further, supporters of the park, particularly the Yosemite Fund, have tried valiantly to make up for low federal appropriations through their own fundraising. Bob Hansen told me that members feel an urgency to support the park because of "flat line funding by Congress."[164] All told, between $75 and $100 million a year is spent to run, protect, and restore Yosemite. That's not nearly enough to begin to make systematic changes to transportation systems. In short, as Dahl stated pointedly, "Congress is not paying us enough to run this park."[165]

A second manifestation continues to be intense litigation. Not surprisingly, given their success in rejecting the Merced River Plan, the plaintiffs in those cases have turned their sights on the VYP. The Friends of Yosemite Valley continued to characterize it as "a development plan, not a restoration plan" throughout the 2000s.[166] The group claimed that more than 90 percent of the funds would go for new construction that would translate to increases in asphalt surfaced areas, widening of half the valley roads, a large bus station in Yosemite Village, and new parking lots.[167] In December 2006, the Friends and the Mariposans filed suit to reject the VYP, arguing that it had been based on the 1980 GMP, which had not been implemented, and a river plan that had been rejected.

A recent ruling suggests that the courts will continue to be important actors in any Yosemite plans. In March 2008, the Ninth Circuit Court of Appeals upheld the district court's 2006 ruling on the river plan, finding that the plan "does not describe an actual level of visitor use that will not adversely impact the Merced's Outstanding Remarkable Values . . . because the VERP framework is reactionary and requires a response only after degradation has already occurred."[168] The courts thus stopped several construction projects until a new plan is written. That result was "really surprising" to NPS officials, especially since eight major environmental groups, including the Sierra Club and the Wilderness Society, had filed an amicus brief in support of the agency.[169]

The impact of this litigation has been substantial. Jay Watson, a key player in developing and defending the YVP, warned that "we are rapidly approaching a point where a historic opportunity will become a lost opportunity."[170] Referring to the 2000 Valley Plan, Bob Hansen said that the court ruling "throws everything out."[171] Linda Dahl admitted that the impact had been severe: "Since the lawsuit," she said, "we've hardly been able to do anything." But she rejected the suggestion to appeal to the Supreme Court, pointing out that "we've already burned up $18 million on this."[172]

Staying Relevant

Wider public support as manifested through either Congress or the courts is unlikely any time soon if for no other reason than the continued lack of a unified, effective pro-change coalition. When I asked one veteran NPS official (who asked to remain anonymous) why the debates over Yosemite policies are so intense, he answered simply, "People just love this place, so they dig in their heels on what they want." The legal actions are illustrative. Perhaps the one thing that the various litigants have in common is a stubborn, emotional attachment to the park. Because of that, although many environmental groups actually supported the NPS plans, the actions of two smaller, less mainstream, organizations inspired a 2008 article in the *Economist* to conclude that "conservationists," broadly defined, are blocking renovation efforts for fear of greater development in the park.[173] Dahl, who had also been involved in developing the restoration plan for the Everglades, reiterated the sense of emotional attachment that can make rational discussion and potential collaboration challenging. When I asked her to compare the situations in the two parks, she responded, "The level of debate at Everglades is so much more sophisticated down there. For example, they all have dueling models, but at least they're sharing ideas. Here it's street-fighting with uninformed sharp objects."[174]

These political obstacles have fostered two policy results. One is that cars are still prevalent in the valley. In July 2008, I mentioned to an interpretative ranger who has been in Yosemite for nearly a decade that it seemed like there were fewer cars in the valley. "Not really," he answered. "Many days you'll come here and find wall-to-wall cars." Dahl echoed his comment: "If you're here on weekends, there are too many cars in the valley."[175] She added that on six occasions in the summer of 2007, the agency had to turn cars back because the park was too crowded. When I asked her whether she thought the NPS would ever get cars out of the valley, her answer was simple and direct: "No."

The second policy result is an alternative strategy, one that involves the incremental changes that Radanovich and others have endorsed. Park personnel hope that they can use that approach to improve natural conditions in Yosemite. As one ranger said, "if people don't want to give up driving their car into the park, we can't really stop them. We can't reduce the number of cars, so we try to go about reducing their impact by different routes, like encouraging them to ride the shuttle." The NPS thus is pursuing a lot of "small fixes" with the intention of making overall progress.

This approach should not be dismissed as trivial. Some of the "small fixes" are quite impressive. With the help of the Yosemite Fund, for example, the NPS has completely revamped access, restrooms, and other amenities at Lower Yosemite Falls. Whereas visitors once hustled up a crumbling cement path to the falls to escape the exhaust fumes of a dozen tour buses lined up in the parking lot, today the lot is a restored meadow with a modern restroom facility. Or, they can walk the trail across another meadow just east of the falls. Citing the work there as an example, Bob Hansen pointed to the passage in the 1980 GMP that mandates "markedly reduce impact" and then said proudly, "We have markedly reduced impact."[176] In addition, the agency removed two obsolete hydroelectric dams from the Merced River in 2004; one, the Cascades Diversion Dam, had affected the river since 1917. Removal required two months of demolition work and $2.8 million. Representatives of the Sierra Club and even the Friends of Yosemite Valley applauded the efforts of the agency to restore the natural flow of the river.[177] Further examples are the nearly 300 lodge rooms and more than 300 campsites (in the Lower Pines, Lower River, and Upper River campgrounds) that have not been replaced since the floods. More generally, the agency continues to develop its use of adaptive management—for instance, experimenting with prescribed burns and systematically monitoring the results.

Settling for small fixes is understandable. Agency officials walk a fine line. The visitation statistics in figure 1-1 reflect a trend of substantial concern. Visitation at Yosemite dropped in 1997, in part at least because of the closure from the floods, and the level has stayed down since. In fact, as other figures in this volume show, visitation is down at Yosemite more than in the park system in general or at comparable jewels like Yellowstone. Today, most of the people visiting the park are either relatively affluent white adults or tourists from other nations. Observers ranging from scientists to the editors of the *Economist* have noted the declining interest among young Americans in outdoor activities such as visiting parks.[178] The 2008 article in the *Economist* described Yosemite as especially affected by these trends.[179]

One reason for the decreased visitation at Yosemite is that quite a few people have the perception that getting in is difficult. Many who care about the park worry that continued efforts to restrict automobile access may push the NPS over the fine line where it begins to lose vital support from some important constituents. As noted scholar William Cronon has said, people can learn to love national parks from their cars and that remains a "bulwark" for the defense of the parks.[180] Dahl echoed the concerns about the park being vital to all groups in society and stressed the need to continue to make it accessible. "Who's going to vote for this park in the future?" she asked. "It's important to recognize the changing face of America. The parks are worried about staying relevant."[181]

Those who want to restore Yosemite will have to build a coalition that can foster needed public support for repairing damage from past policies that made it too accessible while keeping it accessible enough. In doing so, the NPS will need some help. Dahl says that the agency "needs to figure out how to flex."[182] For example, perhaps it will continue to limit but not eliminate day parking in the valley. If visitors want to use parking, they will have to get up early and get into the park or ride the bus. "The future will be about choices."[183] Until people are willing to recognize and accept those choices and not just insist on their own version of paradise, we'll see the art of the possible in making repairs to Yosemite Valley.

Love Hurts

Whatever apprehension I had felt coming into the valley, born out of a fear of traffic and crowds and smog, had disappeared during our stay there. We could still smell the ponderosas. We could see the clouds float over the granite towers. We could hear the Merced River. On our last night there, my wife and I walked up toYosemite Falls in the dark, hoping to see one of the moonbows that show up occasionally when the park is bathed in the light of a full moon. I had seen one once years ago when I worked here, but this was Lynn's first visit to the park. When the moonbow didn't appear, I apologized for having gotten her hopes up. She looked at me and said quietly, "Don't worry about it, I'm already enamored with the place."

Falling in love with Yosemite is not hard to do. But people love the park on their own terms, which range from climbing El Capitan to sightseeing from behind the windshield of their automobile. Years ago, talking about national parks in general, some observers worried that we might love our parks to death. Today, in Yosemite, love is not killing the park, but when so many people have dug in their heels on their own version of romance, restoration is a difficult task. As many poets and singers have noted, "Love hurts."[184] At least for the near future, emotions will continue to make rational discussion and dramatic policy changes elusive goals.

The Everglades is a unique ecosystem, a shallow sea of grass flowing very slowly south toward Florida Bay and the Gulf of Mexico.

4

RESTORING WATER TO THE EVERGLADES

A century after man first started to dominate the Everglades, that progress has stumbled. Consequences have started to catch up. It is, perhaps, an opportunity. The great wet wilderness of South Florida need not be degraded to a permanent state of mediocrity. If the people will it, if they enforce their will on the managers of Florida's future, the Everglades can be restored to nature's design.

—MARJORY STONEMAN DOUGLAS AND RANDY LEE LOFTIS, 1988

We have to start getting some progress on the ground.

—STUART APPELBAUM, DEPUTY FOR RESTORATION MANAGEMENT, U.S. ARMY CORPS OF ENGINEERS, 2008

Our canoe bounced in the choppy waves at the point where Indian Key Pass meets the Gulf of Mexico. The trip out had not been difficult, other than a bit of a fight with the rising tide just after we launched in the morning and the need to maneuver to avoid getting swamped by the occasional passing speedboat. But now the afternoon clouds had turned into fog and our visibility was diminishing. As anyone who has canoed in the Everglades knows, it can be a challenge to try to identify individual keys—islands consisting largely of mangrove stands—from the water, but doing so in reduced visibility is even tougher. So, there we sat, studying our chart, trying to figure out how to find Picnic Key, where we could camp for the night.

In the Everglades, such moments of confusion are not that unusual. It is a complex place. Since the days of the earliest explorers, just getting around in this ecosystem has been a challenge, and getting lost, even if briefly, is pretty common. Drawing an analogy between exploring the Everglades and trying to repair them is not a stretch. The complexities and challenges of even a simple canoe trip like ours are just a taste of those facing crusaders like conservationist Marjory Stoneman Douglas and engineers like Stuart Appelbaum. Appelbaum's comment above, made several years into one of the largest restorations of a natural environment ever attempted, suggests that restoring "nature's design" will not be easy.

Water and Grass

When someone is asked to imagine a national park, the image that comes to mind is likely to involve mountain vistas, deep forests, and cascading waterfalls. All of these are found at the great national parks—Glacier, Great Smoky Mountains, Yellowstone, Yosemite, and others. None are found in the Everglades. The Everglades is an ecosystem whose beauty pictures alone cannot convey; to do justice to the subtle splendor of the place, visitors must use words. In fact, some scholars suggest that the Everglades was the first national park created to preserve ecological resources rather than natural vistas.[1] Yet few who have been there would deny that it is not only a unique but a remarkable ecosystem. The vast expanses are just as daunting in their own way as the highest mountain range. Though forests, even trees, are rare, the park is home to an immense array of plant and animal life. As for water, well, it may not be cascading, but it is a river. At the beginning of her classic account, Douglas refers to the "grass and water that is the meaning and the central fact of the Everglades of Florida. . . . It is a river of grass."[2] Given a chance, the water here could constitute one of the great river systems on earth. However, water and grass are useful for other endeavors, which are, unfortunately, completely inconsistent with the preservation of natural places. One does not have to be in the Everglades long to realize that this is a severely damaged ecosystem.

Revising Nature's Grand Design

Originally, the marshes of the Everglades region covered abut 10,000 square kilometers. Even after more than half of those marshes disappeared, much of it under farms and subdivisions, the Everglades is still the largest wetland in the United States, constituting about 1.4 million acres. The Everglades is also the largest subtropical wilderness in the

United States, home to both saltwater and freshwater habitats that endow it with a degree of biodiversity that exists nowhere else in the world. Furthermore, the Everglades is the only place in the United States—and one of only three in the world—recognized as an International Biosphere Reserve, a World Heritage site, and a Wetland of International Importance.[3]

Those attributes were rarely appreciated, however, in the eighteenth and nineteenth centuries. Instead, most early explorers of south Florida condemned the area as a place of endless swamps and voracious insects. Pioneering efforts to settle the Everglades almost always ended badly. One late-nineteenth-century visitor concluded that "the Everglades will always retain its present state of wildness, and thus furnish a safe retreat for game animals, where they will multiply and increase in spite of the advance of civilization."[4] What was that "present state?" The original Everglades consisted of a roughly forty-mile-wide wetland whose waters flowed very slowly south, rarely interrupted by dry land, from where the meandering Kissimmee River emptied into Lake Okeechobee all the way to Florida Bay. When people ventured into this area, they usually did so to exploit the natural splendor, in particular by shooting tropical birds for plumes for women's hats.

Some prescient individuals recognized early in the twentieth century that the Everglades warranted at least some protection. The National Committee of Audubon Societies hired a warden in 1902 to address the bird hunting situation, only to have him murdered in 1905.[5] In 1903, President Teddy Roosevelt established Pelican Island as the first U.S. national wildlife refuge. In 1916, some wealthy conservationists convinced the state legislature to create Florida's first state park at Royal Palm. In 1923, National Park Service director Stephen Mather recommended creation of an Everglades National Park. Congress eventually authorized the park in 1934, and, after years in which congressional opponents refused to provide any funds, finally established Everglades National Park in 1947. The park itself contains roughly the southern fifth (about 21 percent) of what had once been a much larger Everglades ecosystem. However admirable the early efforts, they did little to stop the larger forces that would dramatically change the Everglades.

In the twentieth century, Americans moved to try to "control" nature in even the most daunting places. The Everglades were no exception. In Florida, the U.S. Army Corps of Engineers led the charge. Congress established the Corps in 1802 to assist with navigation and transportation on the nation's rivers. They expanded the agency's responsibilities in 1824 with the first rivers and harbors bill, calling on the Corps to improve navigation on the Mississippi and Ohio rivers. After that, the Corps grew to become an organizational powerhouse with jurisdiction over many of the nation's waterways as well as responsibility for coastal protection and development of wetlands. Congress was instrumental in its growth, because members were eager to support Corps projects in their home states and districts whether or not the projects made economic or ecological sense.[6] In trying to re-engineer the Everglades, the Corps found a project that has kept them busy to the present day.

In south Florida, controlling nature meant doing something about the water. Water, such a precious commodity in so many places today, was historically considered a nuisance in the Everglades, and officials plotted ways to drain the swamps or at least control the water flow. The sheet of water, moving from north to south, ebbed and flowed, creating thousands of small pools and marshes that were great as natural habitat but useless for building farms or houses. Since at least the early nineteenth century, people eager to use the area for even seemingly benign purposes such as bird-watching discussed draining the "swampland."[7] After the massive floods on the Mississippi River in 1927, President Herbert Hoover and the Corps renewed the efforts to subdue natural forces, particularly flooding of areas near waterways. A subsequent flood of Lake Okeechobee, at the northern end of the Everglades, in 1928 prompted a seminal response in the region. Spending $20 million over the next decade, the Corps built the Hoover Dike, a four-story barricade of the waters that had flowed south into the Everglades. The amount of land available for farming in south Florida suddenly increased dramatically, and sugar companies in particular moved in to take advantage.

Efforts to drain, reshape, and develop the Everglades intensified after World War II. A severe drought in 1944–45, heavy rains and flooding in 1947, and then hurricanes in 1947 and 1948 caused millions of dollars of damage to south Florida, leading state officials to ask for federal assistance. One writer described the subsequent effort in these terms: "To prevent future tragedies, man revised nature's grand design."[8] Specifically, the Corps pursued the Central and South Florida (C&SF) Project for Flood Control and Other Purposes as "one of the most massive and complex public works projects in the nation's history."[9] The agency promised that the C&SF project would control flooding, enable development, and also preserve natural resources; ultimately, however, it was an "engine for economic growth."[10] Voters in Florida approved the project overwhelmingly—even the conservationist Douglas supported it—and Congress authorized the project with the Flood Control Act of 1948 and provided the Corps $200 million to make it happen. The Corps devised an unprecedented series of canals, levees, and impoundments to manage the area's water. The water was drained into "railroad-straight canals that allot every drop to irrigation, drinking, or the Everglades—in that order."[11]

A Struggle for Survival

Structural engineering of the Everglades dramatically affected the ecology of south Florida. As historian Michael Grunwald points out, the Everglades were in ecological trouble even before the C&SF project was launched, but by the time the Corps completed the project in the 1960s, the entire region had been completely altered.[12] By the early 1970s, one observer described the end result of all the engineering as "a struggle for survival of the Everglades."[13]

The modifications were extensive in scope. The Corps controlled water flows with 1,400 miles of canals and levees, hundreds of gates and spillways, and more than a dozen massive pumping stations.[14] It turned the restless, 103-mile-long Kissimmee River into a tame, 56-mile-long canal, nearly bereft of adjacent wetlands and thus available to cattle ranchers. Journalists subsequently referred to the Kissimmee as "a channelized and polluted ditch."[15] By the time the Corps finished with Lake Okeechobee, now constrained by the Hoover Dike, they had separated what had been the slow-moving Everglades River from its headwaters by a 20-mile-wide barricade.[16] The project also produced the massive, 700,000-acre Everglades Agricultural Area (EAA) in what had been wetlands between Lake Okeechobee to the north and the Everglades to the south.

With the water under control, population growth and agricultural development exploded. Between 1940 and 1964, the state's population increased from 1.9 million to 6.8 million, nearly four times the national rate. The number of people per square mile jumped from 35 to 126.[17] The newcomers wanted houses, roads, and shopping centers, all in places where they had not existed before. Much of the growth occurred in or near what had been the Everglades ecosystem. The population in the area grew exponentially, from less than 30,000 at the start of the century to more than 5 million at the end.[18] That occurred despite the fact that, as Douglas later noted, "South Florida is probably the worst place on earth to put millions of people."[19] Agricultural activity increased dramatically. The amount of farmland in the state nearly doubled between 1940 and 1964, from 8.3 million to 15.4 million acres.[20] While not all that growth occurred in south Florida, nearly all of the northern Everglades were replaced by sugarcane fields. Between 1950 and 1994, the amount of land in south Florida devoted to sugarcane cultivation multiplied twelvefold, to 435,000 acres.[21] Overall, by the early 1990s, about 27 percent of the original Everglades had been converted to agriculture.[22]

The ecological impacts of those changes have been severe. Decades of building canals, dams, and locks fragmented and damaged the flow, quantity, and quality of water in the ecosystem.[23] The canals carried water out of the system even while usage by the growing population and agricultural interests increased. In wet seasons, the system removed too much fresh water from the ecosystem too quickly, thus leaving insufficient amounts for the dry months. In some years, more than 1.5 billion gallons of fresh water was lost to the ocean.[24] Because farmers and developers had higher-priority claims on the system's water, that the park generally received flows only when it was already inundated.[25] Historical annual flows of more than 450 billion gallons of water had been cut to only about 260 billion gallons.[26]

Intensive agricultural practices also affected water quality. Huge pumps delivered water to the EAA during dry seasons and removed it during the wet periods. The water that flows through the system is con-

Intensive engineering of fresh water flows through canals such as this one and the production of sugar, as at the Domino refinery in the background, have had a severe impact on the Everglades.

taminated by nutrients used for growing sugarcane, such as phosphorous and nitrogen, and other contaminants that previously existed only in low amounts. Farm runoff flowed into both Lake Okeechobee and the Everglades marshes, affecting the chemical composition of the water and thereby causing eutrophication and other problems. Phosphorous, for instance, fed the growth of cattails, which replaced the sawgrass that had dominated the historical sea of grass.[27] Nutrients also have killed certain kinds of algae that had been a crucial source of water in dry seasons for the eggs of aquatic animals.

The impacts on biodiversity have been tragic. Marshes and other bird habitats were replaced by farm fields and housing developments. Development had diminished the rookeries of herons, egrets, and storks by 95 percent of what they had been in the 1930s, when the area was home to tens of thousands of birds.[28] By the 1990s, the park supported less than 10 percent of the nesting birds that it had in the 1930s.[29] Alterations to one part of the natural system inevitably affected others. Today, sixty-eight plant and animal species are listed as endangered or threatened, including everything from crocodiles to panthers.[30]

The changes to the Everglades from development, pollution, and water management did not go unnoticed. Ironically enough, the year 1947 saw not only the floods that inspired the massive C&SF project but also the publication of Douglas's *River of Grass,* which sounded an early warning about the effects of the changes in the environment. By the early 1970s, journalists and observers bemoaned the devastation of what once had been a remarkable ecosystem. The *National Geographic,* for example, published a lead article in 1972 entitled "The Imperiled Everglades," calling the situation in South Florida a "crisis."[31] One national parks historian described the Everglades as being "in a precarious position, plagued by insidious water problems."[32] Later assessments of the massive engineering that had taken place were no kinder. Decades after completion of the C&SF project, many in the media damned it, with one editorial in the *New York Times,* for instance, referring to it as "one of the most colossal environmental mistakes in American history."[33] One environmental history of the region concludes that the engineering projects "effectively killed the Everglades."[34]

Fledgling Efforts to Avoid Losing a Unique Wilderness

Even while the Everglades was suffering, south Florida was not immune to the larger environmental awakening occurring in the United States in the late 1960s and early 1970s. Indeed, the factors often cited for contributing to that awakening throughout the country were magnified in the Everglades, and the seeds for a coalition that would advocate changes to policies affecting the Everglades were sown.

During the late 1960s the activism of environmental groups throughout the country began to increase. A controversy that contributed significantly to the mobilization of Everglades restoration groups at that time involved a proposal to build an airport in the Big Cypress Swamp. The Everglades jetport would be four times as large as Miami International, and to the chagrin of conservation groups, it would be located only six miles upstream of part of the national park. Despite the fact that locating the jetport there would eliminate the flow of water into that part of the park, the Dade County Port Authority nearly ignored the park in its analyses, preferring instead to concentrate on the estimated 50 million passengers who would fly into the area each year.[35] Proponents of the jetport obtained many of the needed permits and began building runways. Environmental activists mobilized to stop the proposal, using insider access to federal government officials to prompt them to raise official questions and media publicity to raise external pressure. In 1969, for example, prominent mystery writer John D. MacDonald warned that "unless the world's biggest jetport is stopped, we will lose a unique wilderness."[36]

Ultimately, the environmental impact statement requirement for the jetport concluded that it would devastate the region's ecosystem.[37] That conclusion along with public pressure caused President Richard Nixon to abort the project in 1970. The administration allowed the runway already constructed to be used for three years but called for it to be closed as soon as possible after that. Nixon's own credibility was at stake, given his pronouncement earlier that month of a decade of environmental concern. Shortly thereafter, in 1974, Congress established Big Cypress National Preserve in the area.

The airport controversy and other issues inspired the area's activists. Conservationists, notably Joe Browder and Arthur Marshall, pushed existing groups such as the Audubon Society and the Sierra Club to become more vocal in their concerns about the Everglades and more aggressive in their actions. In 1968, the National Parks Conservation Association and the Audubon Society established the Everglades Coalition, which consists of an alliance of local and national groups dedicated to restoring the Everglades. Marjory Stoneman Douglas, disillusioned by the changes to the ecosystem that she had loved so long, helped form the Friends of the Everglades in 1969. Other national organizations increased their own involvement in the Everglades, through activities ranging from education campaigns to lawsuits. The pro-restoration coalition would ultimately contain forty-one major conservation groups.[38] Consistent with the advocacy coalition framework described in chapter 1, it also attracted at least the informal participation of politicians, agency officials, and academics.

Another factor often associated with a substantial increase in progressive environmental policies in the 1960s was the reapportionment of state legislatures to more closely achieve "one person, one vote" representation following the *Baker* v. *Carr* Supreme Court ruling in 1962. In Florida, as elsewhere, that shifted power from rural, often conservative, areas to urban, typically more liberal, parts of the state. One consequence in Florida was the nurturing of the political careers of progressive politicians such as Democrats Bob Graham, Reubin Askew, and Lawton Chiles as well as Republicans like Claude Roy Kirk.[39] Encouraged by civic-minded advisers such as Nat Reed, this new generation of policymakers endorsed environmental awareness and pushed for state policies that could benefit the restoration effort in the Everglades. For example, the Florida state legislature passed the Land Conservation Act in 1972 to authorize spending for environmentally endangered lands; in the same year, the state officially mandated minimum flows and levels to protect water resources. Table 4-1 lists the major state and federal actions referred to at various points in this chapter.

Federal policymakers were not immune to the growing demand for attention to natural areas. In fact, another nationwide trend involved the increased growth of tourism and outdoor recreation, which gave rise to many efforts to protect the environment from the late 1960s to the 1980s, particularly in natural areas. In south Florida, tourism is one of the two primary industries (the other is agriculture), and it generates billions of dollars of revenue each year.[40] Those involved in the tourism industry were eager to ensure that the Everglades continued to exist as an ecosystem that would attract visitors. Indeed, after a couple of years during the 1960s when no water was delivered to the park, some worried that the park might cease to exist at all. In response, in 1970 Congress required that at least a billion gallons (315,000 acre-feet) of water be sent to the park each year. Still, that "minimum delivery" amounts to only 20 percent of the average annual natural flow, and it is not nearly enough to sustain the ecosystem.[41]

The increase in scientific knowledge about ecological processes was crucial to many environmental efforts throughout the country as well as in south Florida. By the 1980s, many scientists better understood the problems that afflicted the Everglades as well as what could be done to

Table 4-1. *Timeline of Actions Related to Policy Changes at the Everglades*

Year	*Level*	*Title*	*Goal*
1947	Federal	Organic Act	Establishes Everglades National Park
1972	State	Water Resources Act	Mandates minimum flows and water levels
1987	State	State Water Management Improvement Act	Calls for district plans to clean up water systems
1992	Federal	Water Resources Development Act	Authorizes restoration of Kissimmee River
1992	State	Consent decree	Formalizes water quality goals
1994	State	Everglades Forever Act	Calls for water quality improvements
1996	Federal	Water Resources Act	Establishes the South Florida Ecosystem Restoration Task Force
2000	Federal	Water Resources Development Act	Authorizes the CERP
2007	Federal	Water Resources Development Act	Authorizes funds for CERP
2008	State	U.S. Sugar Land Deal	Authorizes purchase of farmland from U.S. Sugar

Sources: National Research Council, *Progress toward Restoring the Everglades* (Washington: National Academies Press, 2007), pp. 34–35; Michael Grunwald, *The Swamp* (New York: Simon & Schuster, 2007).

attempt restoration.[42] Even with increased knowledge, however, scientists recognized that any restoration plans inevitably involved a substantial degree of uncertainty. One of the most intriguing developments involved the use of adaptive management to facilitate government efforts to alter complex ecosystems. Often too succinctly summarized as systematic experimentation, adaptive management encourages resource managers to take action, monitor the consequences, and adjust plans for future endeavors. Some of the most important early applications of adaptive management were developed for the Everglades, and scientists quickly realized that incremental adjustments would not be sufficient. Adaptive management analysts recommended in 1992 that the "modest efforts" and "tinkering" that had occurred to that point would not achieve the scale of restoration required, which could be attained only through a "substantial commitment" by public authorities.[43]

Finally, as in other environmental efforts in the United States, institutional actors were increasingly involved in developing conservation efforts. In many cases, the primary agency pushing for changes was the U.S. Environmental Protection Agency, established in 1970; however, the EPA has typically maintained a "relatively low profile" in the Everglades.[44] Two other federal agencies and one state agency have been much more active. The National Park Service (NPS) and the Fish and Wildlife Service (FWS) have become increasingly strong advocates of protection and restoration of the Everglades ecosystem. At the state level, the Flood Control District assumed responsibility in 1972 for water quality and wetlands; the agency was renamed the South Florida Water Management District (SFWMD) in 1976. The SFWMD operates the water management system to pursue multiple goals, including maintenance of water supply, flood control, and, at least occasionally, environmental protection.

The key institutional player in the Everglades has been, of course, the Army Corps of Engineers. As discussed earlier, the Corps had a history of aggressively pursuing engineering projects, but by the late 1960s, it "found itself in a state of disequilibrium."[45] Economists and ecologists alike were increasingly criticizing the agency for pushing water projects that were both economically and environmentally problematic. Further, the dam-building era that had generated so many Corps projects seemed to be coming to an end, with large multipurpose dams attracting increasingly intense criticism for cost overruns and adverse impacts on ecosystems. The peak year for dam building in the United States was 1960, and by 1979 more than 90 percent of the dams listed by the Corps in its national inventory were completed.[46] However, the Corps, showing itself to be "willing to expand and diversify to protect" itself, changed its decisionmaking procedures to allow more public participation and adopted, at least rhetorically, a new mission: environmental restoration.[47] In the Everglades as elsewhere, that meant potentially undoing many of the things the agency had done in the past.

Inherently Contentious

As promising as all these developments were for advocates of restoration in the Everglades, a consensus for change did not exist in south Florida. The reasons include the nature of what had been lost, the nature of what was replacing it, and the inconsistent behavior of the Corps, an agency that would be crucial to any restoration coalition.

The endangered ecosystem was not one that immediately inspired people's zeal to conserve or restore it. The ecosystem that was disappearing was mostly a bug-infested swamp, apparently so vast that losing another acre or two to a farm or housing development did not seem especially tragic to many. Further, animal lovers found it harder to embrace a crocodile half-submerged in a swamp than a wolf howling from a mountaintop against the backdrop of a full moon. Visitor totals are not an optimal measure of public affection for a park, but they are a reasonable proxy. Visitation at the Everglades always trails that at the other crown jewels of the national park system. Consider the figures in 1989, for instance, when restoration proponents were trying to build a coalition for change. In that year, the Everglades hosted less than a million visitors (961,000), far behind the number at the Grand Canyon (4.3 million), Yosemite (3.4 million), and Yellowstone (2.7 million).[48] In addition, visitation trends were not promising. As figure 1-1 in chapter 1 shows, visitation to the Everglades dropped significantly in the 1970s and has stayed relatively flat since. Public opinion surveys of residents of Florida have shown some support for the park, but national surveys are less convincing.[49] For instance, in a 1998 nationwide survey conducted for the National Parks Conservation Association, Colorado State University researchers asked respondents to name their favorite national parks. The results, based on 300 responses, showed the three most popular parks to be Yellowstone (65 votes), Yosemite (37 votes), and Grand Canyon (34 votes). The Everglades received only 2 votes.[50] Thus, the early losses of the ecosystem did not inspire overwhelming protest, certainly not at the national level.

In addition, replacing the disappearing ecosystem were two principal components of the American dream: homes and farms. In south Florida, farms often involve sugar. Policymakers have protected and nurtured the U.S. sugar crop with tariffs and subsidies for two centuries, beginning with the imposition of high tariffs on sugar in 1816, largely to protect a few cane growers in Louisiana.[51] In the twentieth century, Congress added import quotas to further support the sugar industry, even at the cost of artificially high prices for U.S. consumers. By 1994, when restoration of the Everglades came to the fore, the General Accounting Office estimated that such programs transferred at least $1.4 billion a year from consumers to producers; today, the amount is closer to $2 billion.[52] Efforts to eliminate or even reduce support for sugar producers have traditionally been either short-lived or unsuccessful due to the fact that the programs have been popular on both sides of the partisan aisle. In 1974, for instance, Congress abolished sugar quotas only to have President Reagan, despite his free trade rhetoric, reimpose them in 1982.[53] Similarly, the Clinton administration, even while advocating restoration of the Everglades, spent millions of dollars purchasing sugar to prop up the industry in the 1990s. Since the early 1930s, many sugar producers have located in south Florida. Today, three counties in south Florida produce more than half of the nation's sugar;[54] approximately two-thirds of the Everglades Agricultural Area is devoted to sugarcane cultivation.[55]

Largely due to government policies and subsidies, sugar production is concentrated in the hands of only a few producers, who are affluent, well connected, and politically influential.[56] The average size of a Florida farm increased between 1940 and 1964 from 134 acres to 380, an increase that is not inconsistent with increases in average farm size in the rest of the country (from 175 to 352) but more exaggerated.[57] More than half of south Florida's sugar production is controlled by two immense corporations, U.S. Sugar and Flo-Sun. Flo-Sun includes several companies, notably Domino and Florida Crystals. The federal government supported these corporations in the 1990s at a rate of more than $100 million a year.[58] In addition to financial support, the companies had priority claims on the region's water supply, and they were able to discharge their own pollutants into the remaining water.

By the early 1990s, sugar growers were making estimated annual profits of $238 per acre,[59] and they were not going to allow massive changes to their traditional use of the Everglades without a fight. One means of fighting involved litigation and countersuits. For instance, they mounted an intense battle against state efforts to reduce pollutant discharges. Another strategy involved financing politicians on both sides of the partisan aisle. The Fanjul brothers, who run Florida Crystals, are a classic example. In 1992, Pepe Fanjul was vice chairman of the Bush-Quayle Finance Committee, while his brother Alfy supported Clinton-Gore and hosted a $120,000 fundraiser.[60] The Fanjuls and other sugar growers have also been very supportive of state politicians, notably Jeb Bush, who served as governor from 1998 to 2007. Sugar growers have also spent millions of dollars fighting campaigns run by Save Our Everglades and other pro-change groups. The battle over a proposed penny-a-pound tax increase on sugar in 1996 prompted Big Sugar to spend $25 to $35 million attacking "environmental extremists," and one of the Fanjul brothers felt compelled to call President Clinton to complain about Vice President Gore's support for the tax, thereby interrupting a conversation the former was having with an intern named Monica Lewinsky.[61] One analyst termed the sugar debate the noisiest and most expensive fight over an environmental issue during that year's elections.[62] The end result was defeat of the tax, with 52 percent of Floridians voting against it.[63]

The opposition to restoration proposals did not emanate from agricultural interests alone; the booming real estate industry also resisted any serious efforts to restore the original wetlands. Florida's population grew significantly in the 1920s and again dramatically after World War II. The state facilitated and encouraged growth by not imposing personal income or inheritance taxes, an obvious attempt to appeal to retirees. That attempt succeeded, and the retirees showed up, looking for houses. As a result, the price of land jumped dramatically.

Such growth has environmental consequences. The real estate boom devoured acres of land and millions of gallons of water a year, and intensive development pumped large amounts of pollution into the remaining water of the Everglades. But as the cost of real estate skyrocketed, the resistance of developers to proposals to return the land to its natural condition increased. For example, the Latin Builders Association became a powerful proponent of the Homestead airport proposal that was presented in the 1990s.[64] In general, developers and real estate professionals have proven adept at influencing policymakers at both the state and federal levels. Studies typically describe them as among the most influential lobbies in Washington. For instance, a 2001 study in *Fortune* magazine listed the National Association of Realtors as the ninth- and the National Association of Home Builders as the eleventh-most-powerful lobby in the country.[65]

The inconsistent behavior of one of the key players in the restoration effort also slowed momentum toward building an effective coalition to repair past policies in the Everglades. While the Corps of Engineers, as described earlier, had gone through some changes in the 1960s and 1970s, any expectations that the agency would pursue a complete reversal of priorities and practices would prove to be unrealistic. Even the scholars who had described the Corps as an agency in transition to greater ecological awareness admitted in their study that engineers still dominated the organization and that "the environmental mandate has not been welcomed with open arms everywhere in the agency."[66] The Corps still showed a willingness to pursue costly pork barrel projects and an eagerness to assist developers. In Florida, the Corps historically approved more than 99 percent of all applications to develop wetlands.[67]

The bottom line was that, well into the 1990s, restoration efforts in the region had not, with the one exception described below, been very effective. Building a pro-change coalition had proven difficult. As the GAO concluded in a study of public participation in Everglades decisionmaking in 1995, "restoration efforts are inherently contentious, and consensus on solutions that directly affect various interests may not be attainable."[68]

Designing a Solution

We had a chart with us, a hint that we were not just making things up on the fly on our trip into the Everglades. No one should attempt to travel by canoe into the area called the Ten Thousand Islands without a lot of water, at least a little bit of skill, and a good chart. Even as my wife and I bounced on the waves in our boat, I felt good about what we had done so far. We had followed the channel markers and made good time on the way out, especially after the tide began to recede. We were sticking to our plan, which we had made in advance. Any endeavor in the Everglades without a good plan was doomed to hardship if not failure.

Making a Dent

Restoration proponents have long advocated careful planning for the Everglades. In 1947, Marjory Stoneman Douglas remarked on a series of Corps reports, "The most important single recommendation of the Everglades Project Reports was for a single plan of development and water control for the whole area."[69] In 1969, John MacDonald made a similar observation: "The Everglades National Park is in such fragile condition that only by the most careful planning could it be nurtured back to health and stability."[70] During the 1970s and 1980s, farsighted activists like Joe Browder and Jim Webb of the Wilderness Society dreamed of plans for more than just incremental improvements.[71] In 1989, NPS scientist John Ogden helped organize a symposium to discuss a "more integrative understanding of the ecosystem."[72]

The Everglades, however, is a complex ecosystem that involves many parts of the state of Florida and many institutional entities. Most proponents of change were therefore realistic that full restoration would be difficult. At one point, after conservationist Nat Reed had noted that "the option is still open to repair [the] system," his colleague Arthur Marshall responded, "Note the word 'repair,' not 'restore.' To restore the Everglades, I'd have to move 300,000 houses."[73] Not surprisingly, then, in the 1990s those seeking to repair the Everglades system focused on individual components of the ecosystem. Those efforts achieved only mixed success. The one fairly successful project showed what could be done to repair the damage from past policies, but other endeavors illustrated the challenges to building an effective pro-change coalition.

The most successful project focused on the Kissimmee River. As mentioned earlier, in the 1960s the Corps had straightened the winding river, which fed Lake Okeechobee, into a much shorter canal called the C-38 and built a series of dams to control water flow. Corps officials had manipulated the benefit-cost analysis to justify the project, which seemed designed only to please local cattle ranchers who wanted to convert adjacent wetlands into pastures and some residents who bought the promise of greater flood control.[74] The ultimate costs of the project included the loss of 70 percent to 80 percent of the basin's wetlands, severe degradation of water quality from fertilizer runoff and cattle waste, and the listing of more than thirty species of fish and wildlife as threatened or endangered.[75] In the larger context of the Everglades, the impact on Lake Okeechobee was severe. In 1988 testimony before the U.S. House of Representatives, EPA officials admitted that, without redress, the lake would die within "the next decade or two."[76] Historian Michael Grunwald called the C-38 a "glorified sewer pipe" and recalled the jokes of fishermen who said that if the lake level was low, they'd ask Orlando residents to flush twice.[77] Local organizations such as the Everglades Coalition and national groups such as the Sierra Club demanded some response.

The repair effort was impressive. The SFWMD began with some experimental attempts to divert canal flow into marshes and other abandoned areas in the 1980s.[78] Using the state work as a blueprint, the federal government got involved in the 1990s. Former governor Bob Graham, now a U.S. senator, worked to encourage a federal response, and in 1992 the Water Resources Development Act was passed, authorizing the Kissimmee River Restoration Project. The Corps, eager to show its adaptability, took on the job of restoring the Kissimmee to a more natural state. Several prominent scientists endorsed the effort, observing that "the Kissimmee River restoration cut a new path scientifically by establishing ecological integrity as the restoration goal."[79] The Corps spent more than $500 million undoing some of the damage that it had done only thirty years earlier, blowing up one of the dams and filling in seven miles of the C-38 canal and thus restoring fourteen miles of the previously crooked river.[80] By 2008, the restoration was about one-third complete and the groundwork had been laid for future progress.[81] Birds, fish, and waterfowl have since returned, along with some of the marshes and wetlands.

Another issue involved efforts to improve water quality throughout the region. In 1987, the state legislature passed the Surface Water Improvement and Management Act, which required the state's water districts to produce plans to clean up their water systems.[82] And in 1988, Dexter Lehtinen, the acting U.S. attorney for south Florida, filed suit against the SFWMD for failing to obtain water quality permits for the water being pumped into the EAA. Battle lines quickly formed. Many environmentalists as well as agency officials from the NPS and the FWS allied with Lehtinen. The SFWMD was supported by the state as well as the agricultural community. After years of polarized debate, the two sides eventually negotiated a settlement in the summer of 1991.[83] The federal government and the SFWMD agreed to pursue water quality improvements in the Everglades, an agreement formalized as a consent decree (see table 4-1). The battle showed the fragile bonds between important actors in a potential pro-repair coalition. After the battle ended, in some bitterness, change proponents were "more eager than ever to devise a consensus plan to Get the Water Right."[84]

Shortly after taking office, the Clinton administration, led by Interior Secretary Bruce Babbitt, took up the fight to reduce dangerous levels of phosphorous in the Everglades. In some areas, the runoff from sugar fields contained 200 parts per billion of phosphorous despite the fact that many scientists considered the highest acceptable level to be only 10 parts per billion.[85] The sugar industry argued that to achieve that level would require 100,000 acres of filter marshes. Babbitt and the administration of Governor Lawton Chiles launched a series of negotiations with sugar growers. Although sugar corporations had fought previous pollution reduction efforts with dozens of countersuits, the negotiations produced state legislation, initially titled the Marjory Stoneman Douglas Act, that moved toward reducing phosphorous levels. Nevertheless, the legislation was produced only through substantial compromise. The legislation called for more than 40,000 acres of storm water treatment areas—paid for mainly by taxpayers, not sugar growers—but it contained significant loopholes and delayed deadlines. Environmental groups condemned the legislation as designed just to resolve pending litigation that could threaten growers, and Douglas considered it to be so compromised that she asked that her name be removed from the title.[86] Table 4-1 lists the legislation by its new title, the 1994 Everglades Forever Act.

Another repair program involved a crucial area known as Shark Slough. The Shark River Slough is home to the largest remaining contingent of alligators in the park, but changes to the slough have resulted in the loss of more than 50 percent of eggs laid each year.[87] Although the slough also provided an essential source of water to the park, the Corps built two massive floodgates to control the water in the area and then closed them between 1963 and 1965. Corps officials were determined to control flooding to avoid creating problems for the Cuban American community known as the Eight-and-a-Half Square Mile, even if it meant drowning Miccosukee Indian hunting grounds above the gates and starving the park of water below. In the 1989 Everglades Expansion and Protection Act, Congress authorized park acquisition of the Northeast Shark Slough, but the Corps was still unable to distribute water effectively to the ecosystem. Corps management of the area allegedly led to the endangerment of the Cape Sable seaside sparrow, prompting eight environmental groups to sue the agency.[88] The NPS demanded more water downstream of the gates, arguing for a buyout of the 350 homes in the area.[89]

Eventually, Governor Bush stepped in and put on hold any plans to buy out the residents until a more comprehensive plan emerged. In the meantime, the controversy displayed the tenuous nature of the pro-repair coalition, pitting the NPS against the Corps, Indians against Cuban Americans, and environmental groups against others more interested in a compromise solution. Grunwald wondered what would become of any comprehensive, multibillion-dollar effort if such a narrow, $80 million project could get "hopelessly bogged down."[90] The situation remained in limbo even as the Corps and others pushed a larger comprehensive plan, discussed later, at the end of the 1990s.[91]

Restoration of the ecosystem would necessarily involve not just repair of past mistakes but also prevention of new ones. A contentious fight over the proposed Homestead airport also illustrated the lack of solidarity among advocates of Everglades repair. After Hurricane Andrew devastated the land between Biscayne and Everglades national parks in the summer of 1992, developers planned a new commercial airport in the area and recruited the support of the Clinton administration as well as the state's two U.S. senators. Although the airport and the associated development posed substantial threats to the ecology of both parks, a perfunctory environmental review allowed the project to proceed. Some environmental groups, notably the Sierra Club, threatened to sue. Others, notably the Audubon Society, declined to fight the proposal, worried that opposition would make the environmental community seem like anti-growth zealots.[92] The Clinton administration hesitated, caught between wanting to respond to environmental demands but not wanting to cease efforts to help the victims of Hurricane Andrew.[93] As described later, resolution of the issue did not come until the next decade, with especially poor timing for one of the major participants, presidential candidate Al Gore.

With all the different issues simmering, policy change proponents realized that repairing individual parts of the Everglades would not restore the ecosystem. Scientists and others recognized that even successful efforts like the Kissimmee restoration project were just a start. Senator Bob Graham (D-Florida), who had been active in many Everglades battles, said in congressional testimony regarding the Kissimmee project, "We have just begun to make a dent in a life-long effort to restore the Everglades ecosystem."[94] Accordingly, in the 1990s, scientists and restoration advocates increasingly emphasized the need for a collaborative, comprehensive plan. One 1995 analysis stated that need in terms that reflect the importance of an inclusive coalition: "A consortium of Everglades scientists and engineers has concluded that the design of composite (i.e., holistic, system-wide) solutions is needed but that this will require long-term commitments and collaboration among rival factions, the vested economic interests, and the leaders of the various water-related local, state, and federal jurisdictions."[95]

Hugging Alligators

In the second half of the 1990s, those pursuing change made progress in building an effective coalition. Some of that progress took place at the state level. In 1994, Governor Chiles put together a commission of thirty-seven voting members representing business interests, environmental organizations, public interest groups, and agencies to address the "grim vision of South Florida on an unsustainable course."[96] As the different groups debated the question of restoration, the relationship among them improved. One analyst stated that "most of the participants also came to understand the complex Everglades ecosystem in a more holistic way."[97]

The impact of human activity on the Everglades has been substantial, but some species, such as these birds and alligators, hang on tenaciously in this wetlands ecosystem.

Progress also grew out of several different federal initiatives. The Water Resources Development Act of 1992 called on the Corps to develop a restoration plan by 1999. In 1993, the Corps and the SFWMD began what came to be known as the Restudy of the C&SF Project. Also in 1993, the Clinton administration used an interagency agreement to form the South Florida Ecosystem Restoration Task Force in order to coordinate ongoing federal repair efforts in the region. The task force initially consisted of assistant secretaries from several cabinet departments and the EPA. The 1996 Water Resources Development Act expanded the task force to include state, local, and tribal representatives and gave it a formal mandate to formulate plans for restoration. In

1995, the administration directed the Fish and Wildlife Service to develop a comprehensive recovery plan for threatened and endangered species, titling the program the Multi-Species Recovery Plan.[98]

In early 1996, for the first time, the Clinton administration formally endorsed a comprehensive new plan for the Everglades.[99] Vice President Gore announced that the Restudy would be a priority program for the administration. Because 1996 was an election year, the announcement kicked off a vigorous competition between federal politicians to see who could promise more money for Everglades repair. President Clinton and the Republican candidates for their party's nomination, all recognizing the value of support in a politically divided Florida, took turns issuing escalating proposals to convert acres now devoted to sugarcane back to wetlands. Vice President Gore and Speaker of the House Newt Gingrich (R-Georgia) jumped in too, prompting conservationist Joe Browder to comment, "It is the environmental issue in the country where Republicans and Democrats are competing to do the right thing."[100] Or, as Senator Graham said more vividly, "Everybody is grabbing an alligator and wants to hug it."[101]

The deliberations continued throughout the campaigning and after it was over. Congress funded the task force with more than $1 billion. Nevertheless, as of spring 1998, the task force had not yet developed an overall strategic plan. Observers stressed the need for urgency. General Accounting Office analysts, for instance, while recognizing that systematic restoration would require a "complex, long-term effort," strongly encouraged the task force to develop a comprehensive plan soon.[102]

Finally, in October 1998, the Corps released its massive, nine-volume, several-thousand-page draft report of the Restudy. The Corps claimed to have used input from more than thirty agencies and professional disciplines but acknowledged that it was still "a work in progress." The agency promised to hold a dozen public meetings and allow for considerably more input in developing what was to be "the only plan that takes a system-wide look at the southern Florida region."[103] All that money that the politicians had been so eager to spend now had a potential target. The Corps estimated that its reengineering of the C&SF project would cost at least $7.8 billion, the cost to be shared "equally between the federal government and citizens of

Table 4-2. *Components of the Comprehensive Everglades Restoration Plan (CERP)*

Water storage surface reservoirs: eighteen conventional water storage facilities of 180,000 acres located north of Lake Okeechobee and east of the Everglades

Aquifer storage and recovery (ASR): 330 wells built to store water approximately 1,000 feet below ground

In-ground reservoirs: at least two water storage areas in quarries left over from rock-mining operations

Stormwater treatment areas (STAs): at least 35,000 acres of artificial wetlands to filter and treat water before it enters natural wetlands

Seepage management: barriers, levees, and pumps for retrieving lost water, designed to add almost 1 trillion gallons a year to the system

Barrier removal: removal of at least 240 miles of levees and canals to reestablish water flow through system

Rainfall-driven water management: adjustments in water delivery schedules to imitate natural flows

Water conservation: at least two wastewater treatment plants for Miami–Dade County to reuse water for recharging east coast wetlands of Florida.

Sources: National Research Council, *Progress toward Restoring the Everglades* (Washington: National Academies Press, 2007), pp. 35–36; Michael Grunwald, *The Swamp* (New York: Simon & Schuster, 2007), pp. 316–17.

Florida."[104] The massive project, announced by Vice President Gore even before release of the draft report, came to be called the Comprehensive Everglades Restoration Plan (CERP).[105] Table 4-2 outlines the components of the CERP.

The CERP included more than proposals for reengineering projects. It also included elements that affected public perceptions and acceptance of the plan, including the casting of the project as a rescue mission, pledges of economic bounty, a scientific method for fixing things if they went wrong, and promises of interagency cooperation. All the elements were potentially fragile or ephemeral. Indeed, what Jim Webb, activist for the Wilderness Society, said in 1992 could have been said about the Everglades coalition in 1999: "No Florida politician could afford to be against protecting the Everglades, but that support was just as wide and just as shallow as the Everglades system itself."[106]

The Rescue Image

Change proponents and a sympathetic media have long portrayed efforts to repair past damage to the Everglades as more of a rescue mission than anything else. In the second half of the twentieth century, that portrayal became increasingly prevalent. Marjory Stoneman Douglas's fears of the ecosystem being "utterly lost" in 1947 were reiterated in mystery writer MacDonald's warning of a "Last Chance to Save the Everglades" in 1969 and *National Geographic*'s depiction of "The Imperiled Everglades" in 1972.[107] Even a publication that has never been accused of ecological zealotry, *Sports Illustrated,* published a long article in 1981 entitled "There's Trouble in Paradise," declaring that "too many people are demanding too much of the state's fragile land and water system" and warning that "the sad fact is that Florida is going down the tube."[108]

By the 1990s and leading up to approval of the CERP, the rescue metaphor dominated public perceptions. In 1993, the U.N. World Heritage Committee added the Everglades to its list of endangered World Heritage sites. The National Parks Conservation Association (NPCA) began referring to the Everglades as "the most endangered national park in America" and calling for "bold and unprecedented measures."[109] In four separate editorials in 1997, the editors of the *New York Times* frequently used rescue terminology—for example, referring to the plan as "the largest environmental rescue project ever undertaken in this country" and as an "opportunity . . . to save what is left of the Everglades before they disappear altogether" and to one specific action as a "lifeline" and an effort "to save a once-wondrous ecosystem."[110] In July 1999, Vice President Gore delivered the CERP to Congress in a brochure entitled "Rescuing an Endangered Ecosystem: The Plan to Restore America's Everglades." When Senator Bob Smith (R-New Hampshire), the Environment Committee chair, endorsed the CERP at the Everglades Coalition conference in early 2000, he urged people to "put aside partisanship, narrow self-interest, and short-term thinking by saving the Everglades."[111]

The potential problem with such an image is that rescue efforts, as opposed to revitalization or renewal efforts, for instance, are generally short-lived. They last only long enough to pull someone out of a burning building or to save someone from drowning. Politicians were willing, almost unanimously, to use language about saving the Everglades, but they were less specific about establishing institutions that would ensure its long-term recovery. Once they had achieved headlines by promising billions of dollars for Everglades repair, many in the public who may have continued to press for change assumed that the ecosystem had already been "rescued."

Enlarging the Economic Pie

Many of the economic arguments about policy changes to south Florida involved the impact on the agricultural sector, particularly the sugar industry. For example, one economic analysis concluded that eliminating the sugar subsidy would take some 450,000 acres out of cultivation, conveniently the same number of acres that scientists estimated would be needed to use as water storage or treatment areas.[112] To many change advocates, that would be a win-win situation. As mentioned before, not only did the sugar growers use and abuse (by polluting) a great deal of Florida's water, but sugar production was highly concentrated in just a few rich hands. Further, the sugar industry had gained a great deal of notoriety for receiving heavy subsidies from the federal government, so characterizing sugar growers as the villain in the history of the Everglades was not difficult. The editors of the *New York Times,* for instance, referred to "Sugar's Sweet Deal," "Sugar's Latest Everglades Threat," and "Sugar's Sweetheart Deal."[113] Another analysis concluded in 1994 that "the government makes sugar growing profitable with a subsidy that practically mandates the destruction of the Everglades."[114]

In addition to casting sugar growers as the villains, proponents of policy change could use many measures to make the case that repairing the Everglades made economic sense even if it did have an adverse effect on some aspects of agriculture. Those arguments focus on the importance of tourism to the state economy. Comparing the agriculture sector with the tourism sector in terms of the contribution of each to the economy of Florida is difficult for many reasons. For instance, how much income and employment from related fields should be included in calculating each contribution? In 2004, farm sector output totaled $7.6 billion; $2.8 billion in farm income could be added to that. The tourism sector, in contrast, brought the state economy $14.4 billion from international travel alone.[115] Indeed, overstating the importance of tourism to Florida is a challenge. The state is first in the country for international expenditures on travel, second (to California) in travel expenditures in general, and third (behind New York and California) in number of visitors from overseas.[116] The state of Florida estimates that tourism brings in as much as $57 billion to the state economy.[117] Those defending the sugar industry from efforts to reclaim land for ecological repair might argue that tourism in Florida involves much more than the Everglades ecosystem, but so too could one argue that agriculture involves more than just sugar producers. Repair proponents could make the economic argument that a dying Everglades would likely do more to slow tourism than a curtailed sugar sector would hurt agricultural output.

In fact, however, those pushing the CERP did not have to make such arguments. Rather than describing the proposed policy changes as a zero-sum game in which either tourism or agriculture would lose, proponents argued that by getting more water into the system, they could increase benefits to all. In particular, they could promise more financial resources. Money, as analyst Dewitt John said in 1994, "will be the central issue in restoration of the Everglades."[118] That was certainly true in developing support for the CERP in 2000. Time and again, participants in deliberations on the CERP showed a willingness to subdue potentially intense arguments about the economics of large-scale repair by stressing the need to present a united front in order to receive the vast amounts of federal money that had been promised. As Stuart Appelbaum of the Corps of Engineers said, "enlarge the pie. Everybody can get a bigger slice and you avoid the conflict."[119] NPS hydrologist Bob Johnson, director of the South Florida Natural Resources Center, an office of scientists that provides management advice and evaluation of CERP projects, looked back on the CERP in 2008 and recalled, "They promised everything to everybody; all the stakeholders defined what they wanted and then they designed the project to meet everyone's needs."[120]

The Selling Point for Science

The larger context for the scientific discussions surrounding repair of the Everglades involves the value of waterways and wetlands. The simplest definition of a wetland is that it is a transitional area between open water and dry land.[121] It can be said, to summarize the history of U.S. wetlands succinctly, that policymakers were very slow to realize and then explicitly recognize the value of wetlands such as those being lost in south Florida. Wetlands are vital, sponge-like areas that assist in flood control, serve as filters that improve water quality, and provide critical habitat for many species. In fact, nearly half of the endangered species in the United States rely on wetlands at some point during their life cycle. Nevertheless, whereas the continental United States contained approximately 220 million acres of wetlands in 1790, two centuries later less than half, about 100 million, remained.[122] Scientists knew that the wetlands being lost in Florida were precious, but they were less sure of what to do about them.

The views of scientists regarding the likely impact of the CERP on the wetlands of the Everglades were mixed. NPS scientists, whose participation in the development of the Restudy was limited, said that the CERP would subsidize benefits for developers and agriculture interests but that it would not "result in the recovery of a healthy, sustainable ecosystem."[123] They warned that water flows to the park would not increase for decades and that even then the increases would not be enough to help. Another group of prominent ecologists, led by the University of Tennessee's Stuart Pimm, Stanford's Paul Ehrlich, Peter Raven of the Missouri Botanical Gardens, and Harvard's Edward O. Wilson, sent a letter to Interior Secretary Babbitt, encouraging more extensive outside review of the plan. The ecologists criticized the CERP for not reestablishing natural flows, instead leaving the once unbroken sheet of shallow water that constituted the Everglades fragmented and disjointed.[124] Some scientific publications called the CERP "creative . . . plumbing on the grandest scale" but also "untested" and "complicated."[125] Proponents of the CERP used two different arguments to make the case to the broader public and to politicians that the science was actually more supportive. One argument was empirical and the other much more theoretical.

The empirical argument involved the previously described repair of the Kissimmee River, which showed that waterways in the Everglades ecosystem could indeed be fixed. Not only did the Kissimmee project pursue ecological integrity, but by the time of the debate on the CERP, the results showed that success was possible. The results did not, however, necessarily translate into overwhelming support for the massive undertaking outlined in the CERP. The simplest lesson from the Kissimmee project, after all, was that restoration was possible if you just got the water flowing. As the project's lead biologist said, "All we had to do was get out of its way."[126] The CERP was much larger and more complex. But at least the Kissimmee outcome gave support to the idea that comprehensive efforts could restore some degree of ecological quality.

The more theoretical argument that the CERP was scientifically justifiable involved the incorporation of adaptive management practices into the plan. Adaptive management, as its name suggests, promises to be a process that can be used to fix mistakes. The fledgling adaptive management programs mentioned earlier may have achieved "little or no active experimentation," but they did foster developments in monitoring and assessment that provided a scientific foundation for later efforts.[127] More than that, however, adaptive management allowed policymakers to accept the glossing over of disagreements on important specifics with the promise that they could address such issues later on. I was part of one of the NRC panels that did an early assessment of the CERP; on a fact-finding trip to Florida in 2002, panelists met with several key participants in the program. One of the most important Corps officials in the CERP, Stuart Appelbaum, told us that the inclusion of adaptive management in the CERP was the "selling point" for the chairman of the Senate Committee on Environment, Bob Smith, whose support was crucial, and for other politicians.[128]

Overcoming Mistrust

Achieving consistent interagency commitment to restoring the Everglades has always been a daunting task. The scale of the repair job is immense, covering all the waterways leading into and coming out of the southern portion of Florida. Policies on the Everglades have to involve federal, state, and local agencies if for no other reason than that there are different jurisdictions in the area. Coordinating those entities and achieving consensus among policymakers can be Herculean tasks. As one analysis noted in 1994, "the situation in the Everglades is complex . . . and authority to take action is highly fragmented."[129] Another analysis, describing the various institutional actors in the 1990s, observed that "there is a fundamental mistrust among levels of government,"[130] largely because agencies' priorities usually differ. For example, the South Florida Water Management District is a powerful entity, and it is not likely to have the same priorities as the National Park Service or the U.S. Fish and Wildlife Service. Certainly district officials are much more sympathetic to the water needs of local users, including the agricultural sector, than NPS officials.

Even at the federal level, interagency cooperation, though essential, was not immediately likely. The Corps and the NPS have always had different ideas about how to manage ecosystems. As mentioned earlier, the commitment of the Corps to change has not been consistent. In a study of numerous natural resource agencies, Jeanne Clarke and Daniel McCool concluded that the Corps "has striven to embark in new directions while also attempting to maintain its ties with its more traditional supporters."[131] In the Everglades, those traditional supporters remain farmers and developers. While some, including Corps officials, argue that the Corps has become "greener" in the last few decades, many involved with the Everglades are not so sure. According to Michael Grunwald, for example, the Corps remains "unrestrained and unreformed."[132] The Corps continued to show its ambivalence in early debates on the CERP, with some officials strongly committed but others unwilling to embrace the new environmental mission of the agency.[133]

Eventually, the three agencies promised to try to implement the plan. Even while admitting that "the plan is not perfect" and that it would "have to make some midcourse corrections," the Corps expressed strong confidence in its ability to make it work.[134] In addition, both the NPS and the FWS came around to supporting the plan after some changes to the earlier draft.[135] These were not rousing endorsements, and they suggested that the long road to implementation that lay ahead might be rocky.

Something for Everyone

The CERP promised much. The plan was designed, as the Corps said, to "benefit almost everyone living in south Florida."[136] Grunwald agreed, telling me in 2007 that "the park was never the starting point for the plan, the starting point was to have something for everybody."[137] In its final form, the CERP estimated that the project would cost $10.5 billion and require at least three decades to complete. As shown in table 4-2, the original plan involved sixty-eight different projects, among them proposals to build eighteen massive, above-ground reservoirs covering 180,000 acres; two below-ground reservoirs rebuilt from limestone quarries; 330 aquifer storage and recovery (ASR) wells to store water to be used in case of drought; a seepage management barrier to prevent loss of fresh groundwater; wastewater treatment plants; and 35,000 acres of filter marshes to improve water quality. The Corps promised to restore Lake Okeechobee and the Everglades while increasing water supplies for the region. The plan called for increasing the current 260-billion-gallon flow to about 325 billion gallons by 2010 and to more after that.[138] Indeed, the plan called for eventually providing more than 1 million acre-feet of additional water annually, with 70 percent of that designated for environmental purposes.[139] However, the plan also acknowledged the competing uses for the water, as the goal of the authorization was to "restore, preserve, and protect the South Florida ecosystem while providing for other water-related needs of the region, including water supply and flood protection."[140] The Corps would thus not recreate the natural ecosystem but reengineer the "quantity, quality, timing, and distribution" of water to repair the damage from past practices.[141]

Was the larger audience convinced? While there was interest, there also was wariness. Indeed, the General Accounting Office warned that modifications that were likely to be necessary to address the uncertainties in the plan could increase its total cost substantially.[142] The *New York Times,* so supportive of the restoration concept for years, admitted that the plan had "serious flaws," which it blamed on the complexity of the circumstances, and urged the administration to get it right.[143] Even the reactions of environmental groups were mixed. The Audubon Society called the CERP "a cutting-edge project," but the Sierra Club expressed skepticism about the wisdom of some of the planned actions, suspicious that benefits would serve economic interests more than ecological ones.[144] The Everglades Coalition was divided on how to respond, wary of project specifics but reluctant to criticize the administration, which was backing the larger concept of restoration.[145]

While that reception was not overwhelmingly enthusiastic, one key part of the larger audience, political leaders, were still eager to see formal authorization of the CERP. In the 1998 election cycle in Florida, some candidates, for state and federal office, frowned at the price tag, but nearly all expressed at least rhetorical support for restoration, and several counties passed bond initiatives to help provide funding.[146] At the federal level, the Clinton administration worked to tweak the plan to provide more environmental benefits earlier—for example, promising in mid-1999 to deliver an additional 79 billion gallons of water to the national parks. The administration emphasized that restoration was the "overarching purpose" of the plan, and the Corps promised (although the pledge was not legally binding) to "deliver 80 percent of the new water to the natural system."[147] In addition, the Corps revised its schedule for undertaking projects within the plan. In response to criticism that it was moving too slowly, the agency projected that forty-four of the sixty-eight components in the plan would be finished by 2010.[148]

In early 2000, the Clinton administration asked Congress to approve the plan as a framework for restoration of the Everglades. Because of the breadth and complexity of the CERP, some specifics in the plan were "not as detailed as typical Corps feasibility studies."[149] The lack of specificity worried some observers. Further, to the chagrin of environmental activists, the 1999 modifications described above had been questioned. Other interests that had supported the original plan, notably developers and agribusinesses, criticized the proposed changes. Further, many key players in Congress and Governor Bush sided with the economic interests.

Ultimately, two specific provisions made CERP more palatable to the environmental groups and other skeptics. First, Congress formally authorized adaptive management as the technique to be used to address the uncertainties sure to arise in implementation, with the costs for pursuing that approach to be split evenly between the Corps and the state of Florida. That was the first time that adaptive management was explicitly authorized in a civil works project.[150] Second, largely in response to the concerns of Pimm and the other scientists, the legislation mandated periodic review by independent scientific panels. The National Academies of Science, which was to provide expert scientific advice on the project, established the Committee on Restoration of the Greater Everglades Ecosystem to appoint National Research Council (NRC) panels (generally ten to twelve impartial experts) to monitor and study specific aspects of the CERP. Activists perceived the provisions as allowing for correction of deficiencies later on, although one admitted that "the compromises and deal making still leave a bad taste with many of them."[151]

By the time the Senate held a hearing on the proposed legislation in May 2000, mixed feelings were apparent, but many important actors had signed on to the CERP. Many hearing participants expressed eagerness, in spite of the remaining questions, to get authorization, again because of their perception that the ecosystem was dying. Bob Smith, one of the most influential senators and the chair of the Environment Committee, noted that time was of the essence and called the need for action "urgent."[152] Governor Bush said that the Restudy had "had lots of input" and that "it is time to move on."[153] Senator Max Baucus (D-Montana) expressed some skepticism, noting that "a lot of the science is not yet complete on this project" and asking for more clarification of the priorities between economic and environmental interests. The governor responded that the CERP represented a "delicate balance" of interests, all of which were affected, and added defensively, "Our own state laws give primacy to the natural system."[154] Ultimately, even many of the critics, including national environmental organizations like the NPCA, realized that without passage the substantial amounts of federal funds that had been promised would never materialize. Thus, a shaky but diverse coalition of environmental and business interests "recognized the importance of unity in order to receive federal support" and endorsed the plan.[155]

Even while the debate over the CERP was concluding, the Homestead airport controversy dragged on. It strained relations between the Everglades Coalition and the Clinton administration, ultimately putting Vice President Al Gore in a precarious position during the 2000 campaign. Miami–Dade County officials and citizens in the Homestead area, still reeling from Hurricane Andrew, continued to demand development of the airport.[156] When Gore's opponent, Bill Bradley, came out against the airport, Gore refused to take a side, fearful of alienating Hispanics in the area and other pro-growth Floridians.[157] Others in the administration, notably EPA administrator Carol Browner and Secretary of the Interior Babbitt, opposed the project. The Clinton administration finally rejected the airport proposal on January 16, 2001, earning the applause of environmentalists and the dismay of local development groups.[158] Gore's unwillingness to oppose the airport may well have inclined some Florida environmentalists to support Ralph Nader in the 2000 election, thus sending George W. Bush to the White House. If nothing else, the hard feelings over the resolution of the jetport proposal suggested that cracks were bound to appear in the coalition that had supported passage of the CERP.

Formal approval of the CERP was nearly unanimous. Senator Baucus ultimately relented, as did nearly all the other skeptics in Congress. On September 25, 2000, the Senate approved the CERP as part of the larger Water Resources Development Act by a vote of 85-1, the only dissenting vote being that of Senator James Inhofe (R-Oklahoma). The House followed suit on November 3, by a vote of 312-2. President Clinton signed the legislation on December 11, 2000, ironically enough at the same time that the Supreme Court was hearing *Bush* v. *Gore* to decide the outcome of the 2000 presidential election. Clinton declared, "This is a great day. We should all be very proud!"[159]

The nearly unanimous support among politicians reflected the kind of consensus that gives rise to warnings from political scientists about excessive compromises and fragile coalitions. One observer termed it the product of "years of delicate negotiations among the Clinton administration, Florida, Congress, and an unlikely coalition of environmental groups, sugar barons, utility companies, home builders, and the Indian tribes that live on the land."[160] Analysts also recognized that it reflected the need for action in an electoral year for both presidential candidates as well as members of Congress such as Republican Clay Shaw.[161] One account considered the CERP "the best deal planners could get, given political constraints."[162] Indeed, as soon became apparent, pro-change advocates had built a consensus by avoiding the toughest issues, leaving them for the implementation stage.

Implementing the Largest Restoration Ever Undertaken

No plan comes with a guarantee of success. Even our little canoe trip produced numerous surprises that made implementation of our plan more challenging. We had not figured on the northeast winds that contributed to the extremely low water level at low tide, which required a much later start than we had planned. The weather forecast had not anticipated the rain that arrived off and on throughout the day. And we had not planned on the fog that was now enveloping our canoe and shutting off sightlines to any landforms farther than a football field away. These complications, as relatively trivial as they were, couldn't help but remind me of the challenges that have arisen in the implementation of the Comprehensive Everglades Restoration Plan.

More Complicated than Brain Surgery

Many people tried to be positive about the CERP in the first years following passage, in much the same way that the public allows new presidents a short honeymoon immediately after they take office. The politicians who had been involved in passage expressed hope for "the largest ecosystem restoration ever undertaken."[163] The South Florida Ecosystem Restoration Task Force, including representatives of seven federal agencies, two state institutions, two local governments, and two tribes, continued efforts to coordinate planning. Many bureaucrats were at least hopeful. In 2002, Corps official Appelbaum acknowledged the myriad of challenges but said that "the processes are coming together."[164] While still critical of some of the political manipulations that preceded passage, NPS scientist Johnson was guardedly "optimistic."[165] Environmental groups offered positive wishes. In 2001, for instance, the Audubon Society devoted a whole issue of its magazine to

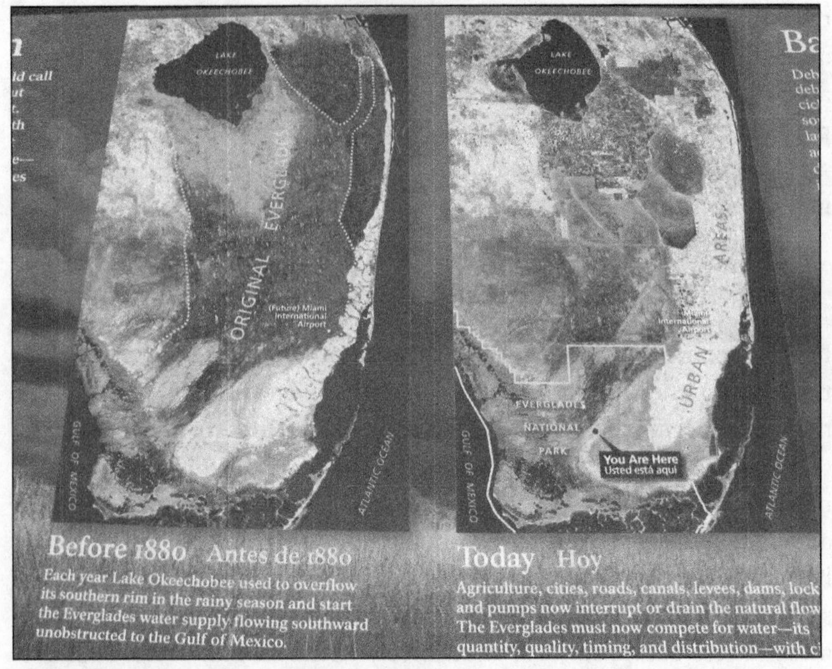
LAKE OKEECHOBEE
ORIGINAL EVERGLADES
(Future) Miami International Airport
GULF OF MEXICO
ATLANTIC OCEAN
Before 1880 Antes de 1880
Each year Lake Okeechobee used to overflow its southern rim in the rainy season and start the Everglades water supply flowing southward unobstructed to the Gulf of Mexico.
LAKE OKEECHOBEE
URBAN AREAS
EVERGLADES NATIONAL PARK
You Are Here
Usted está aquí
GULF OF MEXICO
ATLANTIC OCEAN
Today Hoy
Agriculture, cities, roads, canals, levees, dams, lock
and pumps now interrupt or drain the natural flow
The Everglades must now compete for water—its
quantity, quality, timing, and distribution—with c

As this display shows, changes in the Everglades have been dramatic. The aerial photo on the right shows the interruption of freshwater flows south from Lake Okeechobee and the encroachment of East Coast urban areas on the Everglades.

the CERP, expressing the hope that restoration of the Everglades "will serve as a road map for the rescue of other battered ecosystems across the U.S."[166]

Others were not so optimistic. In a review of the CERP in early 2001, the GAO acknowledged "substantial progress" in formulating the plan but criticized the lack of details "outlining how the restoration will occur."[167] The plan still lacked identifiable starting points, links between planned actions and goals, and specific time frames. Michael Grunwald, who would go on to write the definitive history of the Everglades, was then working for the *Washington Post.* In a compelling series of articles, he warned soon after passage of the CERP that most of the ecological benefits would be "riddled with uncertainties and delays, though it delivers swift and sure economic benefits to Florida homeowners, agribusiness, and developers."[168] Some charged with implementing the plan also were wary. When asked about long-term prospects, John Ogden, a scientist with the South Florida Water Management District, wondered quite candidly before an NRC panel in 2002, "Will we have the will and political momentum later?"[169]

All involved knew that implementing the repairs would be complex. While some liked to say "Just let the water flow," the problems in the Everglades are so extensive that repairs necessarily involve multiple components. When policymakers finally proposed the comprehensive repair effort, it involved numerous stages and various components. Social scientists describe the CERP as at least as complex as other large-scale collaborative resource management efforts, such as the Northwest Power and Conservation Council's fisheries program and the CALFED Bay–Delta Program.[170] The commander of the Corps admitted at the time that the CERP was passed that "the Corps has nothing going on as big and complex as Everglades restoration."[171] Or, as Corps official Appelbaum often said, in more colorful language, "restoring the Everglades was not brain surgery . . . it was much more complicated."[172]

The fragile coalition discussed above elicited sufficient public and political support to lead to approval of the CERP but perhaps not enough to ensure long-term commitment. Lack of the political support needed to overcome the challenges and complexities of implementation over time has been apparent at both the federal and state levels.

President Bush began his term speaking enthusiastically about the Everglades restoration plan, promising on a visit in June 2001 "to restore what has been damaged and to reduce the risk of harm."[173] Even then, though, such rhetoric seemed somewhat hollow. The president's first budget request for the Everglades, $219 million, was an increase over the current year's appropriation but actually smaller than President Clinton's final request and less than many, including Governor Jeb Bush, the president's brother, had hoped for.[174] The events of 9/11 and the Iraq War followed soon afterward, demanding more attention than the Everglades. Throughout his term, the president's leadership on the Everglades was mostly rhetorical. On some related issues, such as banning oil wells from the coast and swamp buggies from Big Cypress, the administration took stands that pleased most environmental groups;[175] on the restoration plan, however, the administration talked a good game but provided little leadership.[176] The extra 79 billion gallons of water that was supposed to go to the Everglades—one of the provisions that bought environmental support of the CERP—was still, as of 2007, "under study."[177] Moreover, in 2007 the Department of the Interior asked the United Nations to remove the Everglades from the World Heritage Sites in Danger list, but many observers considered the move purely political. According to Senator Bill Nelson (D-Florida), such a move would make it even harder to get more federal money because it would diminish the sense of urgency for action. Nelson said of the move, "It's tying my hands behind my back in trying to get the federal share."[178]

Leadership from Congress on the issue has been as inconsistent as that from the White House. Some pro-repair champions in Congress, notably Senator Graham and Senator Smith, had left office or had not been reelected shortly after approval of the CERP. Bob Smith, for example, lost his party's primary in 2002 and moved to Florida, somewhat ironically, to sell real estate. Others did not step up to take their places. At the Everglades Coalition Convention in early 2008, I met with Alcee Hastings (D-Florida), a member of Congress from south Florida who has been involved with the CERP from the outset. In identifying the biggest impediment to implementation, Hastings cited "a lack of coordination of effort and a failure of the federal government to live up to its commitment."[179] He admitted that his own colleagues as well as policymakers at the state level had done little to stem the economic growth in the region, which has had such a strong impact on the repair effort. When I asked him why, he answered simply, "The developers make contributions and politicians respond to that."[180]

The weak support from the White House and Congress was manifest most in the lack of federal money. On our NRC assessment trip in 2002, we heard several speakers complain that they had not been able to do the necessary pilot programs because Washington had been unwilling to provide sufficient funding. Planning for long-term funding, essential to implementation of the CERP, was even more problematic. Appelbaum admitted, "Will Congress lose interest and cut off our funding ten years from now? I don't know."[181] Nick Aumen of the NPS expressed great concern about funding. He said that the current minimal amounts of funding for scientific study would ultimately ensure that the CERP was "not a science-based plan."[182]

Minimal federal funding continued through the 2000s. Until 2007, Congress continually failed to pass a new Water Resources Development Act (WRDA), which would have been the vehicle to contain significant funding for the CERP. In late 2007, Congress finally overrode President Bush's veto of its latest WRDA, thus authorizing some much-needed funds. By that time, however, while the state of Florida had spent about $2 billion on the CERP, the federal government had kicked in only $358 million.[183] That was especially problematic because funding was supposed to be shared in a 50-50 split. As NPS scientist Johnson said, "The state has stepped up but they're incredibly frustrated with federal noncontributions."[184] Appelbaum bemoaned the "slower than desired" pace of federal funding and noted that "it irritates the state greatly."[185] Representative Hastings concluded, "For seven years the federal government did not live up to its commitment and the Everglades paid the price."[186]

State officials moved somewhat faster than their federal counterparts, but they moved toward their own goals. Governor Jeb Bush used his influence to work on several components of the CERP and even announced a $1.5 billion plan primarily to increase water storage capacity just before election day in 2004. Governor Bush also continued to voice public support for the CERP, but he made sure that any specific programs left "plenty of wiggle room to satisfy the needs of constituents like the developers and farmers."[187] In 2003, for example, the state legislature, under pressure from the sugar industry, passed a bill pushing the deadline for meeting phosphorous standards in the water up ten years, from 2006 to 2016.[188] According to at least one source, Governor Bush's administration participated in the "capitulation" to the sugar interests.[189] Another example involved state-owned conservation areas north of the Tamiami Trail. The state generally manages those areas to hold high levels of water so that water is available to supply coastal cities in case of drought even though the deep water creates problems for tree islands and wildlife habitat.[190] NPS scientist Johnson said of state funds, "Funding early on has not been for restoration but for water storage projects."[191] Water storage is essential to restoration, but once water is stored, the park has few guarantees on how it will be distributed.

The lack of consistent state and local enthusiasm for restoration goals has been noticeable. Local political support obviously depends on the views of economic stakeholders. Tom MacVicar, a private consultant to the sugar growers who had previously been a high-ranking official within the SFWMD, told our panel frankly, "We bought into CERP before, but we're not happy about it now. We could solve the water problems for a lot less money with a lower-tech plan."[192] Ogden of the South Florida Water Management District told us in 2002 that the "institutional will, especially among the counties with their emphasis on growth, is problematic."[193] In 2008, Appelbaum told me that because of inconsistent support from state policymakers, the "consensus-building mechanisms may have been lost."[194]

If political support can weaken, it should be able to strengthen again, but participants in the CERP have not expressed optimism. Grunwald wrote in 2007, "The Everglades is still a popular cause. . . . But the enthusiasm of 2000 has faded."[195] Johnson, director of the South Florida Natural Resources Center, concluded quite bluntly in 2008, "There's now a lack of political will to do the restoration parts of the restoration plan."[196]

Beset by Delays

In addition to tenuous political support, tensions between key federal agencies, as anticipated in the earlier discussion of the fragile coalition, surfaced quickly. "Interagency conflicts" and "unresolved tensions" between the Corps and the National Park Service and the Fish and Wildlife Service, glossed over during passage of the CERP, became apparent early on.[197] While relations between the agencies improved somewhat over time, they still had different ideas on how to proceed with the massive repair effort. As NPS official Linda Dahl told me about early efforts to develop and implement the CERP, "they all have dueling models."[198]

One consequence of agency problems in implementing the CERP was the early failure to establish one of the provisions that was fundamental to gaining the political endorsement of several environmental groups before the CERP was passed. The Corps and other policymakers were slow to take the steps necessary to ensure that an effective adaptive management program was in place. In 2003, an NRC panel found that getting a true adaptive management program off the ground would be difficult, concluding that "restoration goals, objectives, and targets for the Everglades are inadequately defined and are not reconciled with the large-scale forces of change in South Florida."[199] Another NRC panel observed that "the investment in science and research relevant to the restoration has eroded measurably within some agencies."[200] The NRC panel on which I served was pleased to find some participants in the CERP dedicated to the idea of adaptive management, but we were also concerned that it was not yet an integral part of the program: "Adaptive management in the CERP is currently more of a concept than a fully-executed management strategy."[201] Adaptive management is not simply flexibility in decisionmaking; it requires systematic targeting and monitoring. Yet few interim performance measures had been developed for the plan. Moreover, managers had barely considered factors outside of the plan that could affect outcomes. Plans for setting priorities and monitoring implementation were still lacking. Agencies also have been slow to develop procedures for standardized data collection and analysis.

Instead of developing the interagency coordination needed to tackle system-wide issues or to fully integrate the adaptive management component of the program, implementing federal agencies tried to break the CERP down into smaller projects that would permit at least some progress. But progress even on those steps has been slow. Further, what was done early on was of questionable ecological benefit. For instance, the Corps approved mining permits for some large limestone excavations. The agency promised that the mines would eventually be converted to water storage areas but, at least according to some observers, all that it accomplished at first was to consume thousands of acres of wetlands and expose groundwater supplies to contaminants.[202] In 2005, Gary Hardesty, the Everglades restoration chief for the Corps, admitted that the agency had not yet implemented even the pilot projects that were designed to test possible strategies.[203]

The slow pace of implementation drew increasing criticism. Consider some of the conclusions of independent reviews mandated by the CERP. In 2001, an NRC panel reported on the aquifer storage and recovery process, whereby water would be pumped into an aquifer to be stored and used at some later time. The panel studied two pilot ASR projects and concluded that current operations lacked important information on issues such as water quality because a proposed modeling program had not been authorized or funded.[204] A second study of the ASR pilots published a year later found that the pilot programs continued to lack systematic procedures for hypothesis testing and experimentation.[205] Another 2002 study concluded that if the CERP did increase water flows, the changes might produce detrimental effects in Florida Bay. Increasing the discharge into the bay, with the associated higher loads of nitrogen and phosphorous, "could potentially increase the frequency, intensity, and duration of phytoplankton blooms."[206] A 2003 study criticized the great degree of uncertainty regarding the impact of increased water flows on the ridge and slough landscape of South Florida.[207] As one report concluded, "the effort required for all these tasks is daunting."[208]

Independent observers were similarly disillusioned. One of the earliest critics, Michael Grunwald, described the complexities and uncertainties in the plan, concluding that "it's not remotely clear whether the Everglades restoration plan will actually restore the Everglades."[209] In direct contrast to the hopes expressed earlier in *Audubon* magazine about the Everglades providing a model for restoration, Grunwald added that "America's largest effort to restore an entire ecosystem may give ecosystem restoration a bad name."[210] Other analysts characterized the CERP as mostly talk and very little action. In a 2004 editorial, for example, Hodding Carter observed that "despite the enactment four years ago of the federal Everglades Restoration Plan, America's largest wetland is most certainly not being restored."[211] In his book *Stolen Water,* Carter expressed considerable frustration with the lack of progress and ultimately concluded that the one way to repair the ecosystem is to "make the entire historic Everglades a national park."[212]

By the late 2000s, the slow pace of implementation of the CERP continued to draw scathing criticisms in other analyses. In 2007, the Environmental Protection Agency published a comprehensive assessment of water quality in the region in which it noted that phosphorous contamination still exceeded CERP goals in nearly 50 percent of the system, that growth of invasive species (particularly cattails) created substantial problems for water quality, and that periphyton (the moist "soil" of the Everglades) had been substantially impacted..[213] Another study on water quality produced similar results. According to hydrologist Herb Zebuth and NPS scientist Nick Aumen, the water quality in Lake Okeechobee was in "terrible shape" and the surface water treatment areas were "often overloaded."[214] Two other important reviews published in 2007 were also quite critical. In a comprehensive report, the GAO described the CERP as plagued by "significant delays, implementation challenges, and rising costs."[215] In another major study, the NRC concluded that "restoration progress has been uneven and beset by delays."[216] Both the GAO and the NRC reports described all of the CERP projects as still at the design or planning stage, if that far along. Many of those projects were six years behind schedule, and any ecological benefits would not be available for years if not decades to come.[217] In late 2007, restoration chief Gary Hardesty said that construction would not begin on any of the projects until at least 2009.[218] Meanwhile, the Everglades continued to decline.

Fixing the Everglades

Making several major changes could help repair the Everglades, but none of them will be easily achieved. One change involves a historical relationship that is literally built on concrete. Grunwald told me in late 2007, "To fix the Everglades, you need to fix the plan, and to fix the plan you need to fix the Corps, and to fix the Corps you need to fix the relationship between Congress and the Corps."[219] He and others involved in the CERP are realistic about the likelihood of such institutional changes. In the WRDA bill that was finally passed over President Bush's veto in late 2007, the proposed far-reaching reforms to the Corps funding process never made it out of Congress. The behavior of the Corps therefore remains entirely responsive to congressional demands. John Woodley, assistant secretary of the army, emphasized the point that Congress calls the shots: "The Corps can't move on a project without congressional authorization."[220]

A second major change would involve a renewed commitment to federal funding for the repair effort. Any substantial progress will require even more money than was promised in 2000. When asked what has to be done to move the CERP along, Representative Hastings responded, "I hate to come back to money, but getting the federal government to authorize the necessary funds."[221] The 2007 WRDA was a massive bill authorizing more than $23 billion for literally hundreds of water resource projects all around the country, including several billion each for the Everglades and Katrina-ravaged Louisiana. More than a few projects in the WRDA were susceptible to charges of pork barrel spending, thus eliciting a veto from President Bush. Consistent with its long history of protecting water projects, however, in late 2007 Congress dealt Bush the first override of his presidency by resounding bipartisan margins (361-54 in the House and 79-14 in the Senate). In an editorial, the *New York Times* said that the action raised "hope for the Everglades."[222] Many participants at the 2008 Everglades Coalition Conference also took hope from the override, seeing it as a sign of a turnaround in the federal commitment to the CERP. Nevertheless, as realistic observers recognized, the WRDA merely authorizes the money; Congress still has to appropriate specific funds for specific projects. The 2007 WRDA would be just one step in that process.

A third major change would involve increasing the amount of fresh water available to the system. Many of those involved in restoration efforts advocate removing structures and just letting the water flow.[223] In some ways, that would be ideal, but with diminishing water supplies, the option is becoming less realistic. As SFNRC director Johnson says, "We can't let it flow because any new flow creates a new demand. It will always be a highly managed system." He argues that in the future there has to be greater emphasis on conservation and developing alternative sources of water through practices such as desalination.[224] Hastings also urged consideration of creating more fresh water through desalination and cited Israel as a model for dealing with water scarcity.[225] Desalination on such a large scale would obviously be expensive, but it may yet be a tempting option.

A fourth major change requires substantial improvements in water quality as well as quantity. While some involved in the CERP note that the state has made progress on improving water quality, the EPA report cited earlier and the evaluations of independent scientists conclude that those improvements are not nearly enough. Tom Van Lent, senior scientist of the Everglades Foundation, argues that the system needs at least an additional 150,000 to 200,000 acres in the Everglades Agricultural Area for water storage and treatment. The most straightforward approach to that task would be to buy the land and use it for water storage, thereby putting an end to the agricultural activity that impairs water quality. Straightforward, yes, but perhaps not politically realistic. As hydrologist Zebuth said in response to that suggestion, "We're going to have to get serious politically if we're going to solve these problems."[226]

Getting Serious or Maintaining the Status Quo?

At this point, the prospects for "getting serious" politically are not especially promising. As the NRC concluded in 2007, "The changes of the past 10-15 years have made the restoration effort more rather than less difficult in many ways."[227] Many trends in South Florida are working against an improved effort.

The most consequential of those trends—or, as historian Grunwald terms it, "the most daunting threat to the Everglades"—is runaway development and growth.[228] Westward sprawl from Miami–Dade County and southeastward growth from the Naples–Fort Myers areas continue to replace wetlands with booming suburbs and relentless depletion of precious water resources. What makes the growth even worse is that, as Grunwald notes, "CERP is designed to feed south Florida's growth addiction, not to cure it."[229] NPS scientist Johnson made a similar observation. "The constraints on restoration are tied to how much water we have to provide to developed areas," he said, adding that "the real reason we included many of the projects in CERP was that we didn't want to preclude future growth."[230]

In addition, the strength of the economic interests that worked to make the CERP more responsive to their needs than to those of the ecosystem is still evident. Sugar remains powerful politically. For example, in 2008, the North American Free Trade Agreement allowed Mexican sugar to be imported into the United States without duties or quotas. However, rather than allow that to hurt the U.S. sugar industry, Congress agreed to buy the equivalent of 85 percent of domestic consumption, regardless of how much is imported. That will not only cost taxpayers nearly $2 billon a year but also foster continued sugar production in South Florida.[231]

Developers too are unlikely to become more sympathetic to the goals of the CERP as the value of land increases. By late 2007, the state had purchased about half the land needed for restoration, but many crucial areas remain private, and the price tag keeps getting higher. One account reported that the price of an acre of land near the Everglades increased between 2000 and 2007 from $2,000 to nearly $20,000 an acre.[232] As for the federal share needed to help purchase those lands, the NRC is pessimistic. The 2007 review panel cited substantially increased cost estimates for the CERP and concluded that "further delays will add to this increase, particularly because of the escalating cost of real estate in South Florida."[233] The growth and development interests in Florida remain influential. As if to underscore that, Governor Charlie Crist, a fairly progressive Republican, said of the CERP in 2007, "Florida remains committed, but we do have to face facts. We do have some economic challenges."[234]

Weather patterns in recent years also have changed, and not for the good of the Everglades. Whether those changes are due to larger global warming issues, any increase in hurricanes and coastal storms can be devastating to much of the region. Hurricanes Katrina and Wilma in 2005 destroyed much of the Flamingo area in the southern part of the park. Widespread damage to the region is also possible if climate change affects precipitation levels. In the mid-2000s, a prolonged and severe drought affected much of the southeastern United States, making battles over diminishing water supplies even more contentious and prospects for ecological flows even more dubious. One likely consequence of climate change, under many scenarios, is rising sea levels. Given the extremely low elevation of much of the Everglades, much of the region could be under salt water within a century. Park officials admit that a two-foot rise in sea levels could transform 50 percent of the fresh water marsh to a salt water system. They assert that such predictions should not undermine restoration efforts as restored fresh water flows could "create a fresh water barrier, hopefully, and keep the rising seas at bay."[235] If such arguments are not persuasive, however, support for an expensive project may be diminished.

Given such trends, a more likely prospect is for continuing incremental change. In its comprehensive 2007 review, the National Academy of Sciences argued that "an incremental approach using steps that are large enough to provide some restoration benefits and address critical scientific uncertainties" is necessary and desirable given current realities.[236] Many involved in the CERP have grasped at this recommendation as a remedy. For example, Assistant Secretary of the Interior Lynn Scarlett said in 2008 that "we need to adopt the incremental adaptive management recommended by the NAS."[237] And indeed, such an approach might foster at least some progress, which is essential for, if nothing else, the morale of those engaged in the implementation effort.

Still, acceptance of only incremental progress is a far cry from the promises that accompanied approval of the CERP in late 2000. Scientist Johnson argues that using such an approach would make fundamental progress less achievable because "the chunks are all tied together and it's hard to break it up in pieces."[238] And such an approach carries with it inherent dangers. Who will determine the sequencing of the individual projects to be pursued? Appelbaum, who is in charge of the Corps's repair efforts, described the "fascination" that many involved with the CERP now have with the potential for the incremental approach; however, he then described how engineers, ecologists, and developers would all recommend a different sequencing for pursuing such an approach. When I asked him whether politics would determine the sequence, he replied, "That's always in the mix." He then added, "Everyone's for sequencing as long as their project is first."[239] Johnson worried that "we're going to pursue the projects that have the most political support and then everybody will get the impression that it's just about politics now."[240]

The importance of politics to policies affecting the Everglade is nothing new. And since it is the dominant factor, those seeking restoration, or at least repair, will have to play a more effective part in the political arena than their fragile coalition allowed in formulating the CERP. The latest development with great potential for the Everglades illustrates this point. In the summer of 2008, the state of Florida announced the purchase of nearly 300 square miles of farmland from U.S. Sugar, the nation's largest sugarcane producer, just north of the park. If the deal does go through, it will be Florida's largest land acquisition ever and could constitute a significant step toward restoration. The acquired land could add nearly a million acre-feet of water storage capacity to the Everglades water system.[241] Shortly after the deal was announced, state officials acknowledged that they hoped to trade some of the U.S. Sugar land to Florida Crystals for 35,000 acres of land vital to restoration plans. The owners of Florida Crystals, the Fanjul brothers, claim that they are sympathetic to the plans and point to their own use of North America's largest biomass power plant as evidence of their "green" side. Many environmentalists and analysts adopted a wary but hopeful attitude.[242]

As the negotiations over the specifics of the deal continued, that wariness seemed warranted. First, the state will have to come up with the money, at least $1.7 billion dollars. Second, some observers and participants in the Everglades restoration effort expressed doubts that the purchase would really contribute that much to restoration. Indeed, some criticized the deal as being more of a bailout of U.S. Sugar than a major move toward restoration.[243] For one thing, the arrangement enabled the struggling corporation to keep farming for as long as possible, at least seven years, at which point the company could lease the land back at much less than market rates. In addition, the expense of the purchase cut into funds that could be used for other, more focused, restoration projects. Third, and related, the NRC warned in its scathing 2008 review of the CERP that any impact from the deal would not be felt for another decade.[244] Skeptics had to wonder whether this, like the 2000 CERP itself, was more a political promise than an actual commitment to ecosystem repair.

The latest overall assessments of the repair effort—from the same scientists, agency officials, interest groups, and journalists that would have to contribute to an effective coalition for change in the future—remain grim. The 2008 NRC review described again many of the problems discussed in this chapter, including substantial delays, lack of federal funding, and the absence of clear priorities and sequencing of needed actions. The panel warned that managers need to move forward with an urgency that has so far not been apparent. In short, as the panel chair concluded, "the ecosystem will continue to lose some vital parts if CERP continues on its present course."[245] I got a succinct summary of the NPS perspective on the future of the CERP at the 2008 Everglades

Coalition Conference when I asked agency scientist Aumen for his overall assessment of restoration prospects. "I'm not very optimistic," he responded.[246]

Neither are environmental groups. A 2008 edition of *National Parks* magazine, for example, described the pending disappearance of the roseate spoonbill as possibly signaling the collapse of the Everglades ecosystem.[247] Finally, for the media's perspective, consider the following. In marking Earth Day of 2008, *USA Today* published an expert assessment of the "10 great endangered places to see while you still can." Along with such vulnerable areas as South Africa's lion habitat and Australia's Great Barrier Reef, the Everglades was listed and described as "disappearing before our eyes."[248]

It Don't Come Easy

We took a chance and paddled for some nearby keys. Our canoe cut through a small opening between the mangroves; then, with the skies clearing just a little, we could see the water opening up somewhat in front of us. Suddenly the chart made sense. We were pretty sure that we knew where we were, so we took advantage of the temporary improvement in visibility and stroked for an island less than a mile away. As we approached and saw the beautiful sandy beach, the palm trees sheltering a tiny campsite, and a family of dolphins cavorting in the shallow water between us and the beach, we knew we were on our way to our destination on Picnic Key.

The Everglades are like that. Efforts can be rewarded, whether on a small scale, as with our discovery of the campsite, or on a much larger scale, hopefully with the CERP. But as Ringo Starr sang, "It don't come easy."[249] Those who are working so hard to see the ecosystem in South Florida repaired can only hope that many of the people who have not yet committed to making that happen take the time and make the effort to experience the real Everglades. Then perhaps, as Marjory Stoneman Douglas wrote, somewhat presciently in 1947, "How far they will go with the great plan for the whole Everglades will depend entirely on the co-operation of the people of the Everglades and their willingness, at last, to do something intelligent for themselves."[250]

The Colorado River in the Grand Canyon is visible from the Watchtower, on the South Rim.

5
REMOVING IMPEDIMENTS TO RIVER FLOWS AT THE GRAND CANYON

Leave it as it is. You cannot improve on it. The ages have been at work on it, and man can only mar it.

—THEODORE ROOSEVELT, 1903

We'll never get back to full natural conditions in the canyon.

—JAN BALSOM, NATIONAL PARK SERVICE DEPUTY CHIEF OF SCIENCE AND RESOURCE MANAGEMENT, 2008

For a rafter running the Colorado River in the Grand Canyon, the most important thing to know is that the Glen Canyon Dam upstream controls the water. So, just as you have to pay attention to the tides if you want to canoe in the Everglades, you have to keep track of how much water is being released from the dam if you want to raft in the Grand Canyon. As the dam operators release water to increase hydroelectric power—generally twice a day, to accommodate surges in demand in the morning and evening—the water travels downriver in what are called "tides" or "pulses." The importance of that fact became clear to me one day after some friends and I hiked back down to the river from Havasu Falls. The hike had been beautiful and rewarding, the falls at the top of the trail so gorgeous and relaxing that we lingered into the afternoon. By the time we got back down to the rafts, tethered at the mouth of Havasu Creek, a pulse of water from Glen Canyon Dam had reached that part of the Grand Canyon, raising the water level in the Colorado River and the side canyons. We had to wade across Havasu Creek to get back to the rafts, and as we did, someone's tube of suntan lotion fell into the creek.

As it floated toward the river, a conscientious guy named John waded out to retrieve it. When he tried to return to the rafts, the current pushed him out toward the river and noisy Havasu Creek Rapid just below. Seeing that he couldn't move and overestimating my own swimming ability, I dove in and swam out to try to help him back to shore. As I reached him, the current in the Colorado picked us both up and floated us toward the rapid. As recreational experiences go, it was less than ideal. Indeed, I learned later that several people had drowned there and that nobody without a life jacket has ever swum Havasu Creek Rapid alone and survived.[1]

Glen Canyon Dam had obviously affected our trip, but that was not surprising. The dam's operations affect virtually every aspect of the Colorado River as it runs through the Grand Canyon, and they have done so since the Bureau of Reclamation completed the dam in 1963. Despite Roosevelt's warning, people did not leave the canyon as it was. Whether or not the changes "marred" the canyon is a matter of opinion, but since 1963, some people have dreamed of removing or breaching the dam and letting the river flow naturally again. However, the costs of such a radical undertaking and the opposition to it have made that dream elusive. Instead, as the Balsom quote suggests, repair of the Colorado River in the Grand Canyon has consisted of often well-intended but less consequential substitutes for removing the major impediment to natural river flows in the canyon.

The Dam and the Canyon

The second reaction of most people when they see the Grand Canyon for the first time, the first one being sheer awe, is to try to see the river. And why not? After all, the Colorado River created the spectacle in front of them, grinding away for more than 5 million years to form one of the most remarkable sights on earth. Sightseers soon realize that they will need to put forth some effort if they are to actually see the river, which lies a mile below them. They can travel to several places on the rim, such as the Watchtower, or they can hike down into the canyon. Sooner or later, many of them do see it, and most then want to see it closer or swim in it or even raft it. The river draws our attention like a mesmerizing blue ribbon, curling and bending at the bottom of the canyon's magnificent rock formations. What many of those sightseers may not know is that what they're seeing is not really the same river that carved the canyon over those millions of years.

Putting Waters to Use

The Colorado River is as important to the southwest United States as any river is to any region in the world. More than 1,400 miles long and draining an area of roughly 250,000 square miles, the Colorado provides drinking water to more than 25 million people and irrigation water to nearly 4 million acres of farmland.[2] Not surprisingly, the Colorado been the subject of intense controversy, negotiation, and litigation. Indeed, the U.S. Geological Survey (USGS) describes the Colorado as "the most controversial and regulated river in the United States."[3]

The Colorado's true power is demonstrated in the Grand Canyon, where millions of years of the river's work has produced one of the truly awe-inspiring places in the world. A World Heritage site as well as a national park, the Grand Canyon is included on virtually any list of the natural wonders of the world, even those limited to just seven places. As president, Theodore Roosevelt designated 800,000 acres of the canyon area as a National Monument in 1908 to protect it from the possibility of intense development, and Congress established the national park in 1919. The Grand Canyon, Roosevelt said, is the one place that all Americans should see, but it is also the destination, as figure 1-1 shows, of nearly 5 million visitors from all over the world every year. British ambassador James Bryce wrote of the canyon, "Why this deep hole in the ground should inspire more wonder and awe than the loftiest snow mountain or the grandest waterfall I will not attempt to explain, but it does."[4]

Beginning in 1869, adventurers and policymakers and engineers explored, employed, and ultimately impounded the Colorado River over the next 100 years. In that year, Civil War veteran John Wesley Powell led the first known descent of the Colorado through the Grand Canyon, penning along the way a classic account that has inspired river rafters ever since. The river at the time was a maelstrom of warm, muddy, sediment-filled water swirling and boiling in riffles, rapids, and eddies. Powell's most famous entry conveys the drama of the adventure: "We are now ready to start on our way down the Great Unknown. We have but a month's rations remaining. We have an unknown distance yet to run, an unknown river to explore. With some eagerness and some anxiety and some misgiving, we enter the canyon below and are carried along by the swift water."[5] Most of Powell's crew managed to survive, and Powell even came back for a second trip, but for decades afterward most people stayed away, not yet sure how the river might be harnessed for utilitarian purposes.

The lack of formal attention to the Colorado River changed after World War I. Recognizing the rapid population growth in the Southwest and the scarcity of water, Secretary of Commerce Herbert Hoover encouraged the states that used the river's water to form a compact in 1922. The compact estimated annual flows to be about 17.5 million acre-feet and divided that amount among upper-basin states (7.5 million acre-feet to Colorado, New Mexico, Utah, and Wyoming), lower-basin states (7.5 million to Arizona, California, and Nevada), and Mexico (1.5 million). The final million was a bonus to the lower-basin states to secure their approval.[6] The compact had many flaws—not the least being that the 17.5 million figure was based on flows in very wet years—and various states, Indian tribes, and Mexico have been fighting over the precious water ever since.[7] The compact did, however, constitute the centerpiece of what people refer to as the Law of the River, and it did legitimize federal projects that divide, store, and use the water.[8] Table 5-1 summarizes the important dates noted in this chapter.

In many ways, the most seminal of the projects was construction in the 1930s of Hoover Dam, just outside of modern Las Vegas. Hoover Dam is arguably the "most significant structure that has ever been built in the United States."[9] Hoover literally stopped the Colorado, creating a lake one hundred miles long and, more broadly, proving to policy-makers that large dams could be built on just about any river in the world. During the subsequent great era of dam building, more than $21 billion was spent on more than 130 western water projects.[10] Hoping to control the scarce supply of water and to generate hydroelectric power while doing so, dam builders looked longingly at the other canyons of the Colorado Plateau.

Table 5-1. *Timeline of Actions Related to Policy Changes at the Grand Canyon*

1908	Grand Canyon National Monument established
1919	Grand Canyon National Park established
1922	Colorado River Compact
1936	Hoover Dam completed
1963	Glen Canyon Dam completed
1982	Glen Canyon Environmental Studies commence
1992	Grand Canyon Protection Act
1995	Final environmental impact statement on operation of Glen Canyon Dam
1996	Experimental spring high-flow test
1997	Experimental spring high-flow test
2000	Experimental low summer flow
2003	Fluctuating flow experiment (January)
2004	Experimental high-flow test (November)
2008	Experimental high-flow test (March)

Sources: Author's compilation from various sources: U.S. Bureau of Reclamation, *Operation of Glen Canyon Dam: Final Environmental Impact Statement* (Washington: U.S. Department of the Interior, 1995); Jeffrey E. Lovich and Theodore S. Melis, "Lessons from 10 Years of Adaptive Management in Grand Canyon," in *The State of the Colorado River Ecosystem in Grand Canyon,* edited by S. P. Gloss, Jeffrey E. Lovich, and Theodore S. Melis (Reston, Va.: U.S. Geological Survey, 2005).

In the 1950s, Bureau of Reclamation engineers proposed building two large dams, one in Dinosaur National Monument and one below the confluence of the Green and Colorado rivers at Glen Canyon. The environmental groups in existence at the time—which were politically immature compared with today's green lobby—withdrew their opposition to the proposed dam at Glen Canyon in return for the abortion of the Dinosaur dam plan and the accompanying promise to protect Rainbow Bridge National Monument from any lake created by a dam. After Congress authorized construction of the Glen Canyon Dam in 1956, David Brower, then executive director of the Sierra Club, and other environmental leaders realized what was being sacrificed. They also noticed that the new Lake Powell's waters would indeed extend to Rainbow Bridge National Monument. Brower and the Friends of the Earth sued but lost in court.[11] They publicly expressed their regret at having signed on to the deal and vowed to never again compromise on such an issue. Just a few years later, when the bureau proposed building two dams downstream in the Grand Canyon area, at Marble and Bridge Canyons, Brower and other environmentalists mobilized. In criticizing the plan, they used many of the tools that environmentalists today commonly use, such as media publicity to support their cause and economic arguments to undermine specific proposals that they oppose. Many observers have characterized this fight as the coming of age of the environmental movement in the United States.[12] Congress officially rejected the proposal in 1968, prohibiting dams on the river between Hoover and Glen Canyon.

However victorious the environmentalists had been in stopping the Grand Canyon dam proposals, the fact was that Glen Canyon was now submerged under the waters of Lake Powell. The dam stands as a monument to the view that rivers, even those that contribute the water to national parks, can and should be harnessed and utilized. At the bottom of Glen Canyon Dam is a sign that you can read while looking downstream at the water, just released from the depths of Lake Powell, that's bound for the Grand Canyon. The sign reads: "Down the River. From this point, the Colorado River flows 720 miles to the Pacific Ocean at the Gulf of California. On its way, it passes through some of the most spectacular country in the world, and pauses at additional dams where its waters are put to use."

Producing Great Amounts of Power

Glen Canyon Dam (GCD) is one of the largest in the United States, standing 710 feet over the bedrock and containing more than 5 million cubic yards of concrete. The crest of the dam is more than 1,500 feet long and wide enough for a road as well as sidewalks. The project was expensive, with the total cost for the dam and the power plant facilities running approximately $300 million, to be repaid to the government through the sale of electricity. When Congress authorized the GCD in 1956, the Colorado River Storage Project Act ordered the secretary of the interior, working through the Bureau of Reclamation, to operate and maintain the structure "for the purposes, among others, of regulating

The Bureau of Reclamation has operated the Glen Canyon Dam since 1963 in order to generate hydroelectric power. The dam controls the flow of water into the Grand Canyon.

the flow . . . storing water . . . [and] providing for the reclamation of arid and semiarid land, for the control of floods, and for the generation of hydroelectric power" and ultimately for producing "the greatest practicable amount of power and energy that can be sold at firm power and energy rates."[13]

The bureau took its mandate to heart, particularly the part about generating hydroelectric power. Hydropower dams are flexible providers of power that can be turned on and off easily by letting more or less water through the turbines in order to generate power at prime (peak or expensive) times of day and reduce power generation at other times. At GCD, the bureau increased daily fluctuations of flows to as much as 30,000 cubic feet per second (cfs) in order to maximize peak hydropower. Environmental groups warned of significant impacts from the resulting unnatural "tides" and sued in court to stop the flows but lost. As a result, the bureau was able to continue to operate the dam the same way through the 1980s.[14] The benefits, at least measured in material terms, were impressive. The maximum combined capacity of the eight electric generators inside the GCD is roughly 1.3 million kilowatts. The dam thus provides substantial amounts of electricity, most of it going to California, Nevada, and Arizona.

The GCD has had significant impacts on the upstream and downstream ecosystems. Upstream, Lake Powell, when full, is 186 miles long and contains, including side canyons, nearly 2,000 miles of shoreline. Not only was Glen Canyon submerged by the "new" Lake Powell, but so too were many grottoes and side canyons, some containing waterfalls and Native American artifacts. In contrast, Lake Powell has become a primary tourist destination. Created in 1972, Glen Canyon National Recreation Area was attracting nearly 3.5 million visitors annually by 1989, only about 20 percent less than the Grand Canyon downstream.[15] Many visitors rent houseboats or just take one-day boat trips to impressive spots within the area, such as Rainbow Bridge. The town of Page, Arizona, exists almost entirely to serve the tourist traffic. Nearby Wahweap Marina is one of the largest marinas and lodging facilities in the region. Other facilities are found at Bullfrog Basin, Halls Crossing, and Hite.

Downstream, many of the rapids that challenged Powell and his men are still there, but otherwise the GCD has substituted its own "river" for the wild Colorado. What had been a warm, silt-laden, seasonally fluctuating river is now a cold, clear, daily fluctuating generator of electricity for the Southwest. Water temperatures that once averaged at least 70 degrees in the summertime are now typically around 46 degrees. Warm-water fish species migrated downstream to survive, or they died off. By the early 1990s, of eight native fish species, three had disappeared, two (the humpback chub and the razorback sucker) were endangered, and one (flannelmouth sucker) was a candidate for federal listing as endangered.[16] In response, federal and state fish and wildlife agencies introduced cold-water species, and the Bureau of Reclamation proudly stated, "In the channel below the dam, rainbow trout thrive in the clear, cold water."[17]

Another major impact involves sediment. Before the GCD, the Colorado moved thousands of tons of silt. In some wet years, the river carried more than 100 million tons of sediment. The gauging station at Phantom Ranch (a small cluster of facilities at the bottom of the Grand

Canyon) recorded a pre-dam average of 380,000 tons a day. By the 1970s, the figure was down to 40,000 per day.[18] The bureau claimed that the dam "has improved water quality in the river by causing sediment to settle";[19] other studies, however, show that water quality has suffered because the nutrients that attach to sediment are left behind in Lake Powell.[20] There are few river runners, having witnessed beach habitat in the canyon disappearing, who would not also question such a statement. In addition, studies have shown that without the protection of sediment, erosion is destroying not just habitat but also archeological sites.[21]

There have been other significant impacts. Not surprisingly, for instance, the vegetation below the GCD in the canyon was much different after completion of the dam. When the level of the Colorado River varied dramatically by season, as it had done naturally, the spring floods scoured vegetation along the banks of the river to the point that only short-lived grasses survived. But by the 1990s, shrubs, willows, and the invasive tamarisk, a nonnative shrub or small tree, had spread along existing shorelines and beaches.

As illustrated by the ongoing story of our raft trip in this chapter, the GCD also greatly affected whitewater recreation in the Grand Canyon. One can argue that because the dam has moderated seasonal flows, river-running companies can operate for a longer period than they could have if the water levels were as low as they historically had been in the summer. Historian Rod Nash calculates that only about 1,500 people traveled on the Colorado River in the Grand Canyon between Powell's exploration and completion of the GCD in 1963. In the fifteen years after that, the number averaged more than 10,000 per year.[22] All of the increase cannot be attributed to the dam, but the dam did lengthen the season. Today, however, the variation in flows occurs daily.

As mentioned earlier, the impact of "tides" from the GCD on river running is profound. Rafters have to time their arrival at some of the toughest rapids in order to hit them when the water level is conducive to running. At low tide, some canyon rapids are nearly impossible to run. Rafters argue that dam operations have made some rapids, notably Crystal, even deadlier than they were before the dam.[23] In addition, beaches are smaller and dirtier, the water is much colder, and driftwood for fires is harder to find. Nonetheless, such problems have not diminished interest. By the mid-2000s, the NPS was allowing more than 26,000 people to float the Colorado each year, 70 percent of them on commercial trips. Those applying for private rafting permits often had to wait about twenty years for their turn.[24]

The simplest conclusion regarding the overall impact of the GCD is to say that it has replaced one ecosystem in the Grand Canyon with another. The changes are fundamental. What had been a wild, free-flowing river is now a managed resource. That fact haunts those who would prefer that one of the world's most famous national parks be more natural. Demands to change or even remove the Glen Canyon Dam have been around since it was built. Some of the earliest advocates for removal found creative ways to express their desires. In Edward Abbey's novel *The Monkey Wrench Gang,* a pipedream of the protagonist featured demolition of the dam,[25] and in 1980, the radical environmental group Earth First! protested the dam by unfurling a 100-meter-long banner that looked like a long crack on the structure's face.

Such demands for removal were generally regarded as extreme, however. As dramatic as some protests were, those who advocated removal showed little skill or even the inclination to build a broad coalition to pursue such changes. By the mid-1980s, Rod Nash asserted that there was "a growing clientele" for preservation of wilderness values in the Colorado River watershed. He argued for protection of "all the remaining free-flowing sections of the river system" but stopped short of calling for removal of the GCD.[26] Indeed, the 1986 Colorado River Symposium at which Nash spoke made eleven substantive recommendations for improving the natural conditions along the river, such as quantifying Native American water rights and establishing a river basin commission, but dam removal was not one of them.[27]

Conflicting Directives

Momentum for a change in policy on the Colorado increased in the 1980s, although for less than obvious reasons. In the early part of the decade, the Bureau of Reclamation pursued the idea of increasing the peak power of the GCD by adding one or more generators. What the proposal did generate was "adverse public reaction."[28] The superintendent of Grand Canyon National Park warned that "if you take the worst possible case of what is being planned, the bottom of the canyon could be devastated."[29] By mid-1981, even James Watt, President Reagan's secretary of the interior, was backpedaling from the expansion idea, but only so far. Watt, already under serious fire for other questionable policy decisions, announced that the bureau would not build the new generators, admitting that "at issue is a key environmental concern—the integrity of the Colorado River downstream from the dam as it flows through Grand Canyon National Park."[30] Watt's comment had to be taken with a grain of salt, however, because the bureau still planned to increase the peak power of the GCD by upgrading the existing turbines. In 1982 the Interior Department commissioned a $50 million series of interagency scientific experiments that Watt and bureau officials hoped would produce data showing that a full-scale environmental impact assessment was not necessary.[31] Until the experiments were complete, they agreed to limit the maximum release of water from the dam to 31,500 cfs instead of the 33,200 cfs that would have been possible with the upgrade.[32]

The experiments did anything but show that further study and discussion were not necessary. Bureau scientist Dave Wegner coordinated the Glen Canyon Environmental Studies (GCES), an interagency effort involving federal bureaucrats, Native Americans, scientists, and private citizens. Wegner and the other scientists struggled to provide objective analyses in the face of what were, in his words, "often conflicting directions."[33] Whatever the Bureau of Reclamation's preferences, the studies showed substantial impacts of dam operations on biological processes, sedimentation, hydrology, and recreation both above and below the GCD. As discussed later, Wegner went on to become an outspoken critic of traditional GCD operations.

The increased questioning of the bureau's management of the GCD received a boost from natural processes in 1983. After record snowfalls in the Rockies that year, the Colorado River experienced several floods, including some in the Grand Canyon when the bureau released massive amounts of water to draw down reservoirs. The floods caused severe problems for rafters on the river at the time and drew criticism from public figures such as the governor of Nevada.[34] During the summer of

1983 alone, huge waves pitched hundreds of rafters into the river, particularly at Crystal Rapid, where they then had to swim for their lives.[35]

To Mitigate Adverse Impacts

Our unplanned swim in the river above Havasu Creek Rapid occurred on the tenth day of a twelve-day raft trip down the Colorado, one of the most spectacular journeys that amateur adventurers can still enjoy anywhere. Like so many others who have taken the trip, we negotiated the big rapids, hiked the side canyons, savored the waterfalls, and enjoyed the sights and sounds of this unique place. It's little wonder that the Grand Canyon can get in people's blood, pulling them back time and again. I can't even remember how many times I've been there. I've hiked down it, up it, and even across it, and I have never failed to be awed. But the Grand Canyon is overwhelming in more than the aesthetic sense. Of the 5,000 or so National Park Service rescues each year throughout the system, typically about 10 percent occur in the canyon. In 1996 alone, for instance, the NPS pursued 482 searches and rescues there but eighteen fatalities still occurred.[36] Most of the rescues are the result of ill-planned hikes, particularly in the summer months, but many involve the river. Now here John and I were, floating toward Havasu Creek Rapid, suddenly in danger of becoming two more fatalities. As we were learning, even with the GCD controlling flows, the Colorado is still a powerful river.

A Mandate for "Alternative Operations"

The momentum for considering policy changes continued to grow through the rest of the 1980s and into the 1990s, involving interest groups, agency officials, and academics in what might have become an effective advocacy coalition. In 1988, three interest groups (Grand Canyon Trust, the National Wildlife Federation, and the Western River Guides Association) sued to demand an environmental impact statement (EIS) on dam operations.[37] State and federal agencies, notably the National Park Service and the U.S. Fish and Wildlife Service, expressed "increasing concern" about the impacts of the GDC on the downstream ecosystem.[38] The National Research Council (NRC) endorsed the GCES and expanded on them.[39] Of most consequence was that the NRC concluded that the studies "proved with increased certainty the need for environmental studies of broader scope."[40] Academics also voiced support for consideration of possible changes. For example, one 1991 study concluded that "the construction and operation of Glen Canyon Dam on the Colorado River has detrimentally affected downstream resource values. These impacts are particularly acute because they threaten the ecological integrity of Grand Canyon National Park."[41]

Federal authorities responded only sporadically during this period. In 1989, Secretary of the Interior Manuel Lujan authorized an EIS on changes to dam operations with the Bureau of Reclamation as the lead agency. To provide data for the EIS, he also authorized a series of "research flows" that would occur over two-week periods in the summers of 1990 and 1991, consisting of three days of steady, 5,000-cfs flows and then eleven days of varying flows.[42] In 1991, in response both to public demand for greater attention to natural values and the fact that the bureau still had not finished the EIS, Congress formally considered action to regulate the flows of water coming from the Glen Canyon Dam in order to protect resources downstream. Senator John McCain (R-Arizona), citing "reports that Grand Canyon resources have been harmed by operations at Glen Canyon Dam," and Representative George Miller (D-California) sponsored the Grand Canyon Protection Act.[43] Despite considerable support, the bill died in the waning days of Congress. Meanwhile, feeling the pressure, the bureau instituted "interim" flows from the dam (maximum release of 20,000 cfs, with daily fluctuations limited to 8,000 cfs) in the summer of 1991.

Congress approved the Grand Canyon Protection Act in 1992. The legislation passed the Senate by a vote of 83-8 and the House by voice vote as part of an omnibus water bill. The act required the secretary of the interior to "protect, mitigate adverse impacts to, and improve the values for which Grand Canyon National Park and Glen Canyon National Recreation Area were established, including, but not limited to natural and cultural resources and visitor use."[44] The legislation also required completion of the EIS by 1994 and called for the process to be paid for by revenue from the sale of hydropower produced by the GCD.[45] Although the legislation explicitly states that it does not supersede the

1956 act authorizing the production of hydropower, most analysts have interpreted it as a mandate to "improve the downstream resources of the Grand Canyon."[46]

The Bureau of Reclamation moved forward on the EIS, finishing the scoping process, providing numerous opportunities for public comment, and producing a draft EIS in 1994. Following more than 33,000 written comments and vast amounts of oral discussion, the final EIS was published in 1995.[47] As required by National Environmental Policy Act guidelines, the EIS contained a range of alternative options for dam operations. The options fit into the three categories: unrestricted fluctuating flows, restricted fluctuating flows, and steady flows. The first category is ideal for hydropower generation; the third category most closely resembles a natural hydrograph. Not surprisingly then, the preferred alternative (the modified low fluctuating flow) was the compromise in the second category. That alternative, which ultimately was adopted, reduced daily fluctuations to the range of 5,000 to 8,000 cfs and provided for "habitat maintenance flows"—simulated floods—to "re-form backwaters and maintain sandbars."[48]

The EIS called for an adaptive management program to implement changes to dam operations. The Glen Canyon Dam Adaptive Management Program (AMP) was to include a planning group, a technical support group, a center for monitoring and research (the MRC), and independent review. The planning group, consisting of twenty-seven representatives of all relevant state and federal agencies as well as stakeholders ranging from environmental groups to Native Americans to water and power interests, was to advise the secretary of the interior on changes to dam operations and subsequent modifications to those changes.[49] The answers to questions regarding such issues as the frequency of simulated floods depended largely on the advice of the AMP.

One provocative possibility was not fully studied in the EIS. The bureau declined to consider dam removal as an alternative:

> Removal of the dam is considered unreasonable in view of: the many established beneficial uses that it now serves, the legal framework that now exists, the investment that the dam represents, the adverse social, economic, and other impacts to the existing human

environment that would result from its removal. Most importantly, Reclamation was directed by the Secretary to evaluate alternative operations for Glen Canyon Dam. The concept of removal is an alternative to operating the dam and, thus, does not address dam operations.[50]

Some policy change proponents responded specifically to the omission in the EIS of the dam removal option. Richard Ingebretsen and other advocates put together a meeting in Salt Lake City of scientists, bureau officials, and environmentalists to discuss this alternative, a meeting that led to formation of the Glen Canyon Institute (GCI) in 1996 as a non-profit group dedicated to removing the dam.[51] In contrast to the bureau's EIS, the GCI, supported by other environmental groups, produced the Citizens' Environmental Assessment of the decommissioning option. This document spells out the arguments for removal of the Glen Canyon Dam, concluding that the dam does not effectively meet its original purpose and is inconsistent with current environmental thought.[52] In addition, in late 1996, the Sierra Club's fifteen directors unanimously endorsed the idea of draining Lake Powell, while leaving the dam standing, by using diversion tunnels to restore natural flows.[53] In 1996 other activists formed a group called Living Rivers, whose mission statement included a call for decommissioning. Removal proponents gained considerable credibility in 1997 when Daniel Beard, former commissioner of the Bureau of Reclamation, endorsed the idea as economically justifiable.[54]

The idea of dam removal (the terms "removal," "decommissioning," and "draining" are used interchangeably) was and remains controversial. The bureau's willingness to ignore that option in the 1995 EIS reflected the inability of proponents to create an effective coalition or the conditions necessary to engage the larger public. Removal advocates have continued to struggle to gain wider support. The proposal has generated conflicting images, inconclusive economic arguments, indefinite science, and a lack of agency commitment.

Breathtaking or Bizarre?

When Glen Canyon Dam was first built, policymakers and engineers described it as an engineering marvel and defined it as a tool to support the rapid growth of the Southwest with cheap hydroelectric power. That image has been under attack since before the dam was completed. Critics have tried to characterize the dam as something quite different, devastating to ecosystems both upstream and down, but they have had only limited success in making that image stick. Water and power interests, Lake Powell recreational users, and many local citizens have contested the critics' view and created a favorable one of their own.

Early arguments about removal focused on the canyons lost underneath the dam's reservoir. As the reservoir that would become Lake Powell filled in behind the dam, some environmentalists and photographers traveled into Glen Canyon to see what would soon be under water. David Brower and photographer Eliot Porter documented the losses in *The Place No One Knew* and condemned the damming as a tragic mistake, though one in which Brower and others had participated.[55] Author Edward Abbey also saw Glen Canyon before it disappeared. In his classic *Desert Solitaire,* Abbey encouraged readers to think about the loss of Glen Canyon under a reservoir by imagining "the Taj Mahal or Chartres Cathedral buried in mud until only the spires remain visible."[56] Those pursuing the image of Glen Canyon Dam as a destroyer of a pristine ecosystem recalled the accounts of pioneers who saw it beforehand, such as John Wesley Powell, who described its "royal arches, mossy alcoves, deep, beautiful glens, and painted grottoes," and Wallace Stegner, who pronounced it "as beautiful a place as exists in the whole canyon country, a Canyon de Chelley with a great river turned into it from wall to wall."[57] In contrast, they derided Lake Powell as "a drag strip for speedboats."[58]

The power of the image of a lost ecosystem has been difficult to sustain. First, those who actually saw Glen Canyon before it disappeared are few in number today. As a result, the lost canyon does not have a strong constituency. Second, to many the lake now has its own kind of beauty; one observer called it "a dazzling blue lake shimmering in a red desert."[59] As mentioned earlier, Lake Powell is the center of the Glen Canyon National Recreation Area, which now attracts several million visitors a year. Third, the town of Page has become an attractive resort destination. People in Page understandably detest the idea of removal; as one Chamber of Commerce official said, "If the lake goes, it would be the end of Page."[60] The town's mayor raised a similar scenario: "You'd have silt, debris, the smell, the bathtub ring around the edge, and Page would become a ghost town."[61] Residents therefore formed an organization, Friends of Lake Powell, to counter the image created by the dam's opponents and thus save the dam and lake. The resulting image of Lake Powell as a prime tourist destination has resonated with many Americans, who prefer more modern comforts to the mud, unpredictability, and warm river water (which would not keep their beer cold as it dragged behind the raft) that might return if the dam is removed.[62]

Those advocating removal of—or at least substantial changes to—Glen Canyon Dam tried to create another image, one of the dam as a cash register, a structure designed to fill the coffers of power companies, which then rewarded authorizing politicians. Abbey suggested the term when he condemned the dam, claiming that "its only justification is the generation of cash through electricity for the indirect subsidy of various real estate speculators, cotton growers, and sugar beet magnates."[63] Colorado River historian Donald Worster noted that even the Bureau of Reclamation was not ashamed of the idea that "running nonstop down in its turbine chamber was a cash register, counting up for tourists the dollars constantly being earned by the sale of electricity."[64] Even relatively impartial historians describe Glen Canyon Dam as a "cash register dam" because its primary purpose is to "bring in revenue by selling electricity to Phoenix and other Southwestern cities."[65] A neutral news source like *U.S. News and World Report* asked, "Should we trade the Grand Canyon's wonder for cheap electricity?"[66] The cash register image is disturbing because it reflects monetary gains that go mainly to hydropower producers and power companies, and the image is reinforced by the impression that such dams exist also to serve the interests of politicians. *Newsweek* reported that a 1997 congressional hearing on the dam removal idea brought out "one of the dirty little secrets of Western dams"—in short, that they exist more for pork barrel reasons than for economic ones.[67]

Defenders of the status quo contest that image as too simplistic, arguing that the GCD provides benefits other than hydroelectric power. An important part of that argument is the dam's role in controlling the water supply. Lake Powell holds about 40 percent of the entire store of water in the Colorado River basin, about 15 million acre-feet that could be distributed to thirsty southwestern communities during a drought.[68] That water is important not just for current usage but also in the allocation of water to the upper- and lower-basin states according to the 1922 compact. Legal scholars have stated that the Law of the River and the 1922 compact are "not a barrier to changes in the operation of Glen Canyon Dam," but they have not been as explicit regarding removal of the dam.[69] Lower-basin state officials consider the dam vital and deplore the idea of removal. Crystal Thompson of the Central Arizona Project once told me that "draining Lake Powell would wreak havoc."[70] The Metropolitan Water District of Southern California, user of 1.2 million acre-feet per year, also opposes the removal option. So too do upper-basin water interests, which argue that if Lake Powell is drained, they will have no guarantee of a dependable source of water or insurance against future needs. One upper-basin official said that if Lake Powell is drained, the area would "have to totally stop growing" to reserve enough water to send downstream to meet the requirements of the compact;[71] however, that may be an overstatement.[72] Although many scientists question the efficiency of the reservoir as a water supply vehicle, defenders of the GCD have succeeded in raising doubt that the dam exists just to benefit the power companies and their political friends.

Opponents of the GCD have also tried to present it as an unsafe structure. Indeed, the dam was embedded in Navajo sandstone, a porous and somewhat crumbly rock. Some warn that the foundation is tenuous.[73] Others warn that even if the dam's foundation holds, the GCD, like a lot of other massive dams, is vulnerable to silt building up in its reservoir to the point that collapse of the dam is possible.[74] Scientists estimate that as much as 65 million tons of sediment settles into Lake Powell each year. As it accumulates, the capacity of the lake is reduced. The sediment already has reduced capacity by about one-fifth. Some say that at current rates, Lake Powell will "become a 186-mile-long mud flat" within 150 to 300 years[75] and that silt could cover the valves that allow for drawing down the reservoir for safety reasons within 100 years.[76] Moreover, the Living Rivers group has presented relatively new models suggesting that the dam could break within sixty years, thus releasing a "500-foot wall of water scouring every living thing out of Grand Canyon."[77]

Defenders of the dam, particularly Bureau of Reclamation engineers, have contested that image as well. They argue that the dam is safe and that it will be for centuries. In addressing the concerns about siltation, bureau engineers argue that sediment in the lake settles as the water slows, thus spreading it out on the bottom rather than piling it up at the base. In 1997 one engineer predicted that at current rates, the lake would not have to be dredged to clear the turbine intake pipes for 500 years.[78]

The image of a crusade is perhaps the most potentially powerful image that proponents of dam removal can cultivate—a chance to make a dramatic statement regarding human relationships with nature. Some argue that such a dramatic action would be a strong symbol of a new attitude in the American West and elsewhere toward natural areas.[79] Former bureau commissioner Beard termed the idea of removing one of the world's largest dams "breathtaking."[80] Sierra Club president Adam Werbach said, "I don't want to be known as part of the generation that killed the Grand Canyon" and went on to call removal "the largest, most important restoration project in human history."[81] GCI president Richard Ingebretsen has argued in different forums and journals that "draining Lake Powell Reservoir is the right thing to do."[82] Further, advocates argue that the crusade to restore rivers throughout the United States was already under way and that removal of Glen Canyon Dam would be only one more, albeit the largest, dam removal in the nation. Indeed, as if to verify that view, the founder of the Friends of Lake Powell worried that "if they get rid of this dam, they would go for the next one, and then the next one."[83]

Critics of removal have been somewhat successful in portraying advocates as not crusaders but misguided zealots at best, duplicitous manipulators of a gullible public at worst. One commentator wrote that "everyone from the Bureau of Reclamation commissioner to the Navajo Nation's natural resources director to house-boating enthusiasts has lined up to denounce the Sierra Club for giving legitimacy—and momentum—to the radical fantasy."[84] Many Arizonans consider the idea "another outlandish example of environmental dementia."[85] One Flagstaff resident who claimed to be sympathetic to environmental issues said of removal advocates, "This makes them look like absolute fools; if you support something ludicrous, it undermines your credibility."[86] One analyst concluded that Sierra Club officials were at least as interested in using dam removal to reclaim their role as a crusading group or as a form of penance for David Brower's having allowed the dam in the first place.[87] When I asked Steve Magnussen, the bureau's director of western water operations, about the removal proposal in 2000, he answered similarly: "My personal view is that it's not a realistic possibility, but it is a good fund-raiser for environmental groups."[88]

Perhaps of most consequence is that most members of Congress also ridiculed the idea of removing the dam or draining the lake. In 1997, members of two House subcommittees held a joint hearing on the removal idea, an event that received front-page coverage in the *New York Times*.[89] The hearing was entitled *The Sierra Club's Proposal to Drain Lake Powell,* a title possibly intended to suggest that the environmental group's position on the dam was an isolated one, shared by few others. Members of the House and Senate from western states expressed "disbelief" and worse at the proposal. Representative James Hansen (R-Utah) termed decommissioning a "bizarre idea."[90] Senator Ben Nighthorse Campbell (R-Colorado) dismissed the idea, saying that it "would be just plain silly to even contemplate it."[91] While Sierra Club president Werbach, Brower, and a few others testified in favor of the idea, representatives from state governments, water and power interests, and recreational businesses condemned it as foolish.[92] Those condemnations had some effect. One indication was that even when the GCI (headquartered in Salt Lake City) and the national Sierra Club were gaining publicity for pushing the removal option, the Utah chapter of the Sierra Club announced that it would not endorse the proposal.[93]

Lack of Economic Tools

The economic arguments about the option of decommissioning the GCD also were ultimately unconvincing to the larger public. Beard, the former commissioner of the Bureau of Reclamation, argued in a *New York Times* editorial that "draining a reservoir and restoring a canyon may just be the cheapest and easiest solution to our river restoration problems."[94] Critics of the idea, however, dismissed that assessment.

Settling the disagreements through economic calculations has proven elusive, given the lack of agreement on even the fundamental issues that would be affected by removing the dam. Estimating the benefits and costs as well as I can, I argue that the economics show that the benefits of maintaining the status quo have been generally overstated but that determining the benefits of decommissioning is more problematic.

The specific points of this argument are summarized in table 5-2. The table shows four possible scenarios. The "traditional operations" column refers to historical operations of the GCD before the 1980s. The "interim operations" column refers to the system in place at the time of the EIS. The "adopted change" column reflects the modified low restricted fluctuating flow alternative recommended by the Bureau of Reclamation in the EIS and ultimately adopted for dam operations. The fourth column, titled "seasonal steady flows," is the closest approximation of costs and benefits if the GCD was removed or breached. Again, the bureau did not provide a systematic assessment of the removal alternative, but it did assess operations under a seasonally adjusted steady flow regime that would include habitat maintenance flows. Of the alternatives examined, that one most closely resembles the natural hydrograph that would exist if the dam were not there.

The differences between the scenarios are readily apparent in the first five rows of the table. Under any operation other than seasonal steady flows, the river is subjected to high releases, low releases, and high daily fluctuations on a substantial percentage of days in the year. The difference is especially noticeable in the fifth row. Under traditional operations of the dam, daily fluctuations (tides) exceeded 6,000 cfs more than 90 percent of the time; that percentage would be zero if the dam were not there. Seasonal steady flows could exceed 20,000 cfs on a substantial number of days, depending on the use of habitat maintenance flows (simulated floods).

One major economic issue involves the potential loss of hydroelectric power. The GCD produces approximately 5,000 gigawatt-hours of power each year, power that is flexible in that it can be generated immediately by changing the rate of release of the water.[95] Calculating the exact number of users of GCD power is difficult because half of the power goes to six large utilities that get power from various sources and then send it to 1.3 million end-users. The other half goes to smaller suppliers, who supply about 400,000 customers. The NRC estimated that hydropower generation from the dam could be valued at roughly $50 to $100 million per year.[96] Table 5-2 shows a summary figure in row 6, the annual cost of lost hydropower based on the contract rate of delivery (in millions of nominal dollars). Several other figures are available, but all show a similar pattern: traditional operations would cost less and seasonal flows (or dam removal) would cost substantially more per year than any other possibility.

The lost power cost is obviously just a summary figure and subject to question. Removal advocates argue that other sources of power, such as the nearby Navajo Generating Station, have much more capacity than the GCD and thus could replace the lost power, but those sources have other issues, notably their contribution to air pollution and climate change, and they have much less flexibility than hydropower. Removal proponents also claim that the dam provides only 2 to 3 percent of the electricity in the region, a loss that could easily be made up by "a few energy saving measures."[97] Beard called the potential loss of hydropower "minimal," especially when compared with the cost of maintaining the dam.[98]

Western Area Power Administration officials warned that, to the contrary, the power lost by removing the dam could be expensive to replace.[99] They and other defenders of the status quo contend that the million or so users of GCD power would bear the costs of the lost electricity and conversion to other sources. One counter argument is that electricity from the dam historically has been heavily subsidized by the federal government, so asking consumers to pay a higher price for it is not that unreasonable.[100] Nevertheless, claiming that benefits from the decommissioning of the GCD would far exceed costs is unrealistic unless the claim is made that decommissioning will also spur the long-awaited conversion to solar and wind and other new energy technologies.

Any benefit-cost analysis must also involve recreation, but any calculation of benefits to recreation from dam removal is subject to intense argument, simply because one vacationer's heaven is another vacationer's hell. Removing the dam might make for a more natural river downstream, but what would happen to the millions of recreational

Table 5-2. *Alternative Operating Scenarios for Glen Canyon Dam*

Scenario	*Traditional operations*	*Interim operations*	*Adopted changes*	*Seasonal steady flows*
Maximum releases	31,500	20,000	25,000 (higher if simulated floods)	18,000 (higher if simulated floods)
Maximum allowable daily fluctuation (cfs)	30,500	8,000	8,000	1,000
Percent of days of releases < 8,000 cfs	90	29	29	< 1
Percent of days of releases > 20,000 cfs	72	19	19	5 to 27
Percent of days of fluctuating releases > 6,000 cfs	97	54	54	0
Annual cost from lost hydropower ($ millions)	0	36	44	124
Whitewater camping	Declining	Minor improvement	Minor improvement	Major improvement
Annual increase in recreational benefits ($ millions)	0	3.7	3.9	4.8
Percent probability of net gain in sediment	41	76	73	100
Harm to cultural resources	Major (336 sites)	Moderate (< 157)	Moderate (< 157)	Moderate (< 157)
Native fish	Stable to declining	Potential minor gain	Potential minor gain	Potential major gain
Endangered fish species	Stable to declining	Potential minor improvement	Potential minor improvement	Potential major improvement

Sources: U.S. Bureau of Reclamation, *Operation of Glen Canyon Dam: Final Environmental Impact Statement* (Washington: U.S. Department of the Interior, 1995), pp. 16, 18, 58–65; U.S. General Accounting Office, "An Assessment of the Environmental Impact Statement on the Operations of the Glen Canyon Dam" (Washington: 1996), p. 123.

users of Lake Powell, who spend more than $400 million a year in the local economy?[101] According to some statistics, the Glen Canyon National Recreation Area is the nation's second-most-popular overnight camping site, behind only Yosemite.[102] As for the downstream river, the managed Colorado River in the Grand Canyon currently hosts more than 20,000 fishermen, roughly 20,000 rafters, and 33,000 day-trip floaters each year.[103] These people invest considerable sums to visit the canyon, more than $20 million annually from rafters alone. A first glance suggests that these numbers would decrease if the dam were removed, the lake emptied, and the downstream river less dependable in terms of amount of flow. Such a conclusion may be premature, however. The same rafting economy that thrives in the Grand Canyon could quickly flourish in a restored Glen Canyon, thus replacing at least to some extent the lost house-boaters.

Imagine the anticipation that would be generated by publicity surrounding the removal of the GCD. Further, no one knows whether more people might prefer to raft a less predictable but more natural Colorado River. Changes to dam operations could improve some aspects of whitewater recreation in the canyon. As table 5-2 shows, for instance, campsites could increase and improve with more sediment in the river and less flushing from daily tides. In purely economic terms, the annual gains, in nominal 1991 dollars, are not overwhelming. A fair conclusion is that the net outcome would be a loss in revenue but perhaps not as much as some in the Lake Powell recreation community have prophesied. It must be kept in mind, however, that this is a purely material calculation. Putting a dollar value on the experience of rafting a free-flowing river is much more difficult.

Some other issues also nearly defy quantification. On the subject of water storage, critics of the GCD argue that Lake Powell is very inefficient, with as much as 8 percent of the water lost to evaporation and bank seepage—roughly 1 million acre-feet per year, an amount significant enough to meet the needs of a major metropolitan area.[104] Indeed, evaporation is the second largest "consumer" of Lake Powell water, behind only agricultural irrigation. One analyst argues that the evaporation losses from a single Labor Day weekend could provide the water supply for 17,000 houses in the West for a year.[105] Further, as sediment fills the lake, the amount lost to evaporation will become a larger and larger percentage of the available supply. While admitting that the lake loses much water to evaporation, defenders of the status quo argue that the water will return to earth somewhere, most likely in the Colorado River watershed.[106] However, at least according to Ingebretsen, that "somewhere" is the Midwest, where the water comes to earth again mostly in the form of rainfall.[107]

Still another obstacle to quantifying costs and benefits from dam removal involves the potential ecological and cultural benefits upstream and downstream. Assigning a dollar value to the endangered fish species, lost archeological sites, and affected vegetation mentioned earlier is more likely to extend than resolve any debates. To complicate matters further, proponents argue that the different ecosystem downstream has had some positive effects for some species, particularly coldwater fish such as trout.[108] Part of the reason this calculation is so elusive is that one has to compare short-term impacts. The GCI has argued persuasively that many of the environmental benefits from removing the dam are "long-term" and thus difficult to estimate fairly.[109]

The bottom line of the economic analyses is that there is no bottom line. Economic arguments do not provide easy answers to the tough questions regarding dam removal. One of the most thorough of the efforts to quantify costs and benefits, by Scott Miller, realistically concluded that we "simply lack the tools necessary to evaluate with any precision some of the costs and benefits of decommissioning the dam."[110]

The Uncertain Science of Jeopardy

The larger scientific and historic context for possible policy changes on the Colorado River is fascinating. During the 1990s and early 2000s, without any coordinated effort from Washington policymakers, scientists and activists all over the United States launched a new era in river management policies. On rivers big and small, they have pursued changes, such as the simulation of natural flows and the reestablishment of aquatic habitat, to attempt to restore the natural conditions that existed before human beings altered them. Those efforts have resulted in a wide range of outcomes, but the most dramatic changes occurred with the removal of dams to recreate free-flowing waterways.[111]

Like the economic analyses, however, the scientific arguments only went so far in terms of supporting removal of the GCD. Scientific analyses showed that the operations of the GCD had impacted the Colorado River and that changes to those operations could benefit the river's ecosystem. The Glen Canyon Environmental Studies previously mentioned, which were published in dozens of technical reports during the mid-1980s, concluded that dam operations could be changed to "minimize losses of some resources and to protect and enhance others."[112] Table 5-2 shows just a few of those benefits. The final four rows of the table show increases in sediment, reductions in loss of cultural and architectural sites from changes to dam operations, and improvements for native and endangered species. Further, while those improvements would be minor under the interim and preferred alternatives, restoring some semblance of the natural hydrograph (possibly through dam removal) could produce major improvements.

The potential increase in sediment in the downstream ecosystem is important. Most scientists recognize that sediment is crucial for maintaining beach habitat and water quality in a river. The NRC concluded in its 1996 review of the GCES that the river below the dam still had adequate amounts of sediment but that the lack of natural flows curtailed effective distribution. For example, "occasional high flows" are necessary to lift sand onto beaches.[113] That sufficient sediment would remain in the downstream river in future years without changes to dam operations is a dubious proposition. Without question, not as much sediment is flowing through the system as in pre-dam years. In addition, if sediment does pile up in Lake Powell, economic benefits such as water storage and recreation will diminish.

On the other side of the ledger, however, the cost of dam removal would increase as more sediment piles up. In short, where would the sediment go? Most analysts agree that draining the lake would involve significant technical difficulties, not the least being what to do with the toxic sediment on the bottom and the bathtub-like ring around what had been Lake Powell. The GCI argues that impounded sediments would "flush quickly" and that the bathtub ring would weather off in as few as ten years.[114] However, the obvious repository for sediment washed downstream is Lake Mead, the reservoir created behind Hoover Dam. Increased evaporative losses in this lake are just one possible additional cost of so much sediment flowing downstream.

Reducing the damage to cultural resources would also be an important benefit of changing dam operations. Most of the cultural resources along the river are managed by the Hualapai tribe and the Navajo Nation, but many of the sites have high spiritual significance to Havasupai, Hopi, Paiute, and Zuni Indians as well.[115] Until the mid-1980s, most analysts did not think that dam operations affected cultural resources because generally they were not found below the high-water mark in the river corridor. However, a study by the NPS and USGS in 1989 showed operations of the GCD contributed to erosion and the loss of necessary sediment deposition at such sites. The GAO concluded in 1996 that archeological sites "have become increasingly exposed to erosion and ultimate destruction."[116] Moderating flows or removing the dam could reduce the damage significantly.

The impact of dam operations on fish, endangered species of fish in particular, also was an important part of the scientific debate. The U.S. Fish and Wildlife Service, the agency responsible for the protection of most species in the United States, issued a final biological opinion in the EIS in 1994. The FWS produced a "no jeopardy" finding regarding potentially affected nonfish species such as eagles and falcons, meaning that proposed changes to operations of the dam were not likely to harm them. However, the FWS opinion did say that the preferred alternative was likely to jeopardize the humpback chub and the razorback sucker.[117] The agency recommended modifying the preferred alternative with seasonally adjusted steady flows about 25 percent of the time to assess the impact of other flow operations on those species. In response, the Bureau of Reclamation promised to use adaptive management to monitor and measure the effects of the chosen alternative and the "research" flows.[118] The FWS still issued a jeopardy opinion on the preferred alternative and expressed its support for the seasonally adjusted steady flow alternative (table 5-2). [119] Bureau officials disagreed with the jeopardy opinion and the FWS recommendation of the seasonal flow alternative, claiming that the argument that "the ecology of the Grand

Canyon will be supported by steady flows is not supported in the document or in the literature."[120] The bureau did agree to continue studying the impacts of different flow regimes.

Scientific arguments for dramatically changing GCD operations or even removing the dam (approximated by seasonal steady flows) were not ultimately persuasive. One major reason was that many of the conclusions were tentative, as suggested by the use of the word "potential" in table 5-2. Scientists could not know what would happen under alternative flow regimes. The impact of the GCD is so consequential that drawing conclusions on what might occur without it or under substantially revised operations is difficult. At the most basic level, only about 16 percent of pre-dam sand levels came from sources other than the mainstream river blocked by the Glen Canyon Dam. FWS scientist Sam Spiller once asked me, "Are you ever going to be able to get to restoration dependent on sediment with just 20 percent of the sediment?"[121]

A second, related, reason was explicit recognition of the uncertainty and the inclusion of the adaptive management program. As early as 1997, for example, the Grand Canyon Trust called for "more scientific information to help us make an informed decision."[122] Whether or not the inclusion of adaptive management in the final plans would produce that information or was simply a disingenuous effort to get plan approval is an issue still argued in GCD debates. For instance, John Weisheit, one of the founders of Living Rivers, says that adaptive management was "a facade from the beginning."[123] On the other hand, many scientists expended an immense amount of effort in the realm of adaptive management in subsequent years. Either way, the promise of an adaptive management program defused many potential criticisms of the policy changes contained in the EIS and made any potential increase in support for the removal option less likely.

A third major reason that scientific arguments supporting dam removal were unpersuasive is that no dam removal had ever occurred on the scale proposed for the GCD. All of the dam removals in the United States during the 1990s and nearly all to date have involved dams that were small compared with the GCD. The most important case, in many ways, involved the removal of the Edwards Dam on Kennebec River in Maine in 1999. The dam, at the time used to provide a small amount of hydropower, had obstructed migratory fish routes for more than 150 years, but in the 1990s a coalition of anglers, environmentalists, and scientists convinced the state and federal governments to mandate removal over the objections of the dam's owners. Within a year after removal, the fish began returning.[124] If construction of the Hoover Dam had proven that even large rivers could be dammed, removal of the Edwards Dam showed that free-flowing conditions could be restored. Removal of the Edwards Dam was big news all over the country, and it even was mentioned in the same context as possible removal of the GCD, but scientists as well as observers dismissed the option for large dams such as that at Glen Canyon, arguing instead that another way that "dammed rivers can gradually be restored is by mimicking the river's natural flow."[125] Indeed, the Edwards Dam had been generating only 3.5 megawatts of power, about one-tenth of 1 percent of the state's power supply and a fraction of the power generated by the GCD.[126] A *Newsweek* article covering the 1997 congressional hearing made that point by concluding, "The debate over Lake Powell is mostly symbolic, but the impending demolition of smaller dams is real."[127]

Agency Cooperation for Show

Any policy involving the Glen Canyon Dam requires a high degree of interagency interaction. The scale of any action involving the Colorado River is immense. The Colorado River, as mentioned earlier, is one of the longest in the world (1,400 miles) as well as one of the most vital in that it brings precious water to a semi-arid region. Consider, as one indication of the range of political entities involved, the different "interested parties" that voiced opinions on the 1995 EIS. The list includes seven federal agencies, ranging from the FWS and the NPS to the Bureau of Reclamation and the Western Area Power Administration; officials from all seven states that use or depend on water from the Colorado River; six different Indian tribes; and dozens of interest groups representing everything from power boaters to environmentalists to fishermen.

As with other long rivers, such as the Missouri, one likely source of tension on the Colorado is between upriver and downriver states. The battles between upper-basin states (Colorado, New Mexico, Utah, and

Wyoming) and lower-basin states (Arizona, California, and Nevada) over Colorado River water are historic. Secretary of Commerce Hoover had to intervene to shape the compact of 1922. Disputes over river water today are nearly as intense as those that preceded the compact, and they are getting more so under the pressure of droughts caused by changes in the climate. People in Arizona still worry, for example, that in the event of water shortages, the deal that the state made to ensure California's support for federal funding of the Central Arizona Project in 1968 will require it to lose all of its allocation of water under the compact before California cuts back even a gallon.[128] Las Vegas gets 90 percent of its water from the Colorado, and officials there are so concerned about possible shortfalls, particularly from droughts, that they are planning a $2 billion pipeline to carry groundwater from the northern part of the state to the city.[129]

Interagency tensions at the federal level also are virtually inevitable. As suggested by the debate between the Bureau of Reclamation and the Fish and Wildlife Service over endangered species, the different mandates of the various federal agencies involved in managing resources lead to suspicion and skepticism. Perhaps the most fundamental tension is that between the NPS and the bureau. While the mission of the NPS is to protect resources, the bureau has had a mandate to develop large facilities for water and power generation since its founding under the 1902 Reclamation Act.[130] Bureau directors such as Floyd Dominy aggressively pursued that mission, enhancing the agency's reputation. As with the Corps of Engineers, when the Bureau of Reclamation claimed to have a new environmental awareness in the 1980s, many remained skeptical. As one NPS official said to me in 1998, "They have a greener face now, but in reality the green face is more for show than for go."[131]

Getting agreement between these different institutional entities on any proposed policy change has been and will continue to be difficult; getting agreement on dam removal is virtually impossible. Even the agencies that one might expect to be sympathetic to the arguments for removal are skeptical. NPS scientist Jan Balsom, a key player in the EIS and subsequent efforts to implement its recommendations, told me, "Removal is beyond what anybody thinks is realistic."[132]

Glen Canyon Dam, in the foreground, and the Navajo Generating Station, in the background, border the town of Page, Arizona, a hotbed of opposition to removal of the Glen Canyon Dam.

Prohibiting an Option

In mandating changes in the mid-1990s in the management of Glen Canyon Dam and thus the Colorado River in the Grand Canyon, Congress opted for alterations rather than removal of the most important structure, telling the Bureau of Reclamation to continue to operate the dam for hydropower generation but also to pursue experimental "research flows."

However, the debate on dam removal has not gone away. Observers of the 1997 congressional hearing on removal did not find it as laughable as Representative James Hansen and other congressional members had hoped. Former bureau commissioner Beard wrote that the proposal made sense to him: "There was no mistaking the intent of the hearing. The Western lawmakers on the panel wanted to use the forum to embarrass those who support restoration of the canyon. It didn't work out that way."[133] Two *Newsweek* reporters had a similar reaction.[134] Dave Wegner, the lead scientist for the bureau's GCES studies but now an advocate for decommissioning the GCD, told me in 1998,

"The only way we'll see restoration of natural conditions is by removing the dam."[135] Many others have since argued for at least systematic discussion of the possibility, but western congressional delegations have remained vehemently opposed to the idea. Starting with the 1999–2000 budget cycle, western lawmakers prohibited, in a rider to an appropriations bill, the use of any funds "to study or implement any plan to drain Lake Powell."[136] Congress has maintained that prohibition since then.

Implementation of a Process

As John and I floated away from the safety of the rafts still tethered at the mouth of Havasu Creek, we had a few alternatives. The first wasn't working. No matter how hard we tried to swim, the current was pulling us out into the Colorado River and toward Havasu Creek Rapid. The second alternative wasn't great, but it improved our chances slightly. As we battled the current, we managed to stay close enough to shore long enough for someone to throw each of us a life preserver. After we strapped them on, we were at least much more fortified for facing a possible ride through the whitewater. The third alternative was to somehow find a rock and hang on until one of the rafts could pick us up. From where we were, bouncing in the cold water of the Colorado River and headed downstream, that looked like a long shot. Our chances of a successful outcome seemed about as remote as those for proponents of restoration of the natural ecosystem downstream of Glen Canyon Dam.

Pursuing Trade-Offs

With the option of dam decommissioning removed from the discussion, policymakers began implementing the decisions reached in the EIS process. Some analysts referred to that as the start of the era of the "Grand Managed Canyon."[137] Implementation of the mandated changes was going to be a challenge. Engaging all the different political entities involved would be difficult, but changing dam operations, which are more multidimensional than one might first suspect, would not be easy either. Once removal of the dam was ruled out, policymakers had to consider changes in maximum flows, minimum flows, daily fluctuations, possible seasonal fluctuations, differences between wet and dry years, and the mandated maintenance flows, which had never been attempted on such a scale before. Scholars characterized the Glen Canyon Dam adaptive management program as "one of the most comprehensive monitoring and evaluation efforts of a single dam in the Unites States—and perhaps the world."[138] As the NRC stated in a 1999 assessment of the new flow regime at Glen Canyon Dam, "the preferred alternative was and is an experiment."[139] However, some aspects of implementation were more readily doable than others.

The most straightforward issue involved setting the level of flows released from the GCD. As table 5-2 shows, the preferred alternative continued to limit daily fluctuations to 8,000 cfs, which had been instituted as the interim flow several years before. That was substantially different from the daily fluctuations that had exceeded 8,000 cfs over 80 percent of the time in the period between 1963 and 1989, when daily fluctuations had actually exceeded 20,000 cfs nearly 60 percent of the time.[140] While that was an improvement for the downstream ecosystem, the limits went only so far. Daily fluctuations would still exceed 6,000 cfs more than half the time, whereas that would rarely if ever occur under a natural flow regime.

The component of the preferred alternative that differentiated it from current operations of the dam and was less straightforward to implement was habitat maintenance flows. The bureau defined habitat maintenance flows as "high, steady releases within power plant capacity," generally set to occur in March to simulate a spring flood.[141] Scientists realized that simply moderating fluctuating flows was not going to stop the erosion of beaches and sandbars.[142] Strong releases of large volumes of water carried the potential to pull sediment out of downstream tributaries and the bottom of the river channel. Some seemingly disparate stakeholders objected to the idea, the upper-basin states out of fear of losing water stores and rafting companies out of concern about the inability to run trips during "flood" periods, but more of them deemed the experimental flows as at least worth a try.

How often were the experimental flows to occur? The EIS specified that frequency would be determined by the adaptive management program but "only" in years in which certain water conditions existed.[143] By the time Bruce Babbitt, the secretary of the interior, signed the

Record of Decision stipulating the operating criteria for the dam in 1997, that language had been modified somewhat to allow for "dam safety purposes."[144] That was a very subtle modification of the language but, at least according to Norm Henderson, the NPS Glen Canyon Dam research coordinator, it ensured a significant role for the Bureau of Reclamation in scheduling such events. In 1998 Henderson told me, "My guess is that the secretary of interior didn't even know the changes had been made."[145] The short story is that the bureau has to approve such an action before it can take place.

Adaptive management, which involves scientific research and monitoring to plan, measure, explain, and adjust management strategies, is fundamental to implementation of any efforts to restore the Colorado River.[146] At the Grand Canyon, the adaptive management program is manifest as "dam-operation experiments hypothesized to achieve downstream ecosystem benefits."[147] The program has several characteristics. First, managers can adjust programs, a flexibility essential for finding innovative solutions involving complex ecosystems. Second, the need for objective scientific assessment is essential and explicit. Third, relevant stakeholders can participate in planning and thus, at least potentially, expensive litigation can be avoided. The two key institutional players for the Glen Canyon Dam AMP are the Monitoring and Research Center (MRC) and the adaptive management work group (AMWG). The AMWG (and the affiliated technical work group) consists of two dozen stakeholders, ranging as mentioned earlier from water and power interests to state and federal agencies to Native American tribes to environmental groups.

While many of the people involved in the AMP worked diligently to institute the program quickly, several issues that became apparent early on have continued to affect implementation. First, funding for the program comes from a source that is hardly unbiased. The budget, in the early years amounting to about $7.5 million per year, derives entirely from the sales of hydropower.[148] In a 1999 review, the NRC noted that this may create "long-term disadvantages."[149] Or, as NPS official Henderson told me more bluntly, "Because money for the program is in the bureau's budget, they can try to control the process."[150] Second, while the MRC scientists went to work immediately after approval of the EIS and developed a long-term strategic plan for the program, the plan was short on specifics. One NRC panel noted in its 1999 review, "Neither the Strategic Plan nor stakeholder groups have articulated a vision for the future state of the Grand Canyon ecosystem."[151] While planners did produce a final draft strategic plan in August 2001, agreement on meaningful specifics within the plan has remained elusive. Third and related, achieving fundamental agreement among such diverse stakeholders has proven difficult. As the NRC realized, adaptive management "will require trade-offs among management objectives favored by different stakeholder groups."[152] After attending numerous meetings over the years and engaging in conversation with many of the participants, I got the sense that, over time, the different stakeholders have developed respect and even sympathy for each other's interests. However, the AMWG has struggled with making tough decisions on contentious issues. Too often, as MRC director Barry Gold said in 2002, stakeholders "looked to the scientists to bridge those differences."[153]

Developing enough consensus within the AMP to take meaningful action has therefore always been a challenge. Members of another NRC review panel in 2002 cited "impressive advances" in planning but warned that the failure to address fundamental questions, including the needed trade-offs between stakeholders, was problematic.[154] The inability to resolve differences between interests has translated to compromise at best, gridlock at worst. One official from the Western Area Power Administration told me in 2001, "The good thing about the GCDAMP is that it reduces the potential for lawsuits. The bad thing about the GCDAMP is that it reduces the potential for lawsuits. At least litigation settles things."[155]

A Decade of Experiments

However difficult it has been to achieve consensus among the diverse stakeholders involved in the AMP, several experiments have been pursued since completion of the EIS. While occasionally these have been dramatic events, their impacts on the ecosystem have been marginal and ephemeral.

The most publicized of the experiments was the first. On March 26, 1996, Secretary of the Interior Babbitt opened the valves on discharge tubes to increase the flow of the river to 45,000 cfs. Dam operators kept the flow at that level for the next seven days. According to "flood" architect Wegner, the 45,000 figure was a compromise among scientists between flows that might stir up sediment and higher flows that might damage already endangered species. While less than half the average amount of pre-dam spring floods, the level was considerably higher than the typical current maximum release.[156] The floods did move sediment. Weeks after the event, Babbitt claimed that as many as one-third more beaches had been formed and that the flood had "worked brilliantly."[157] Scientists with the Bureau of Reclamation and the USGS also cited evidence that sand deposits had increased at test sites in the Grand Canyon.[158] Still, many were only cautiously optimistic. Bureau scientist Wegner warned that "it's a little early to claim success."[159]

The warnings proved prescient. More than 80 percent of the new beaches were gone within a year. Further, the flood failed to establish backwaters for endangered species such as the humpback chub. Wegner was candid, concluding that "it only worked for six months."[160] Other scientists agreed that the positive effects had been overstated. In 1998, NPS researcher Linda Jalbert said that "it did change some, but the changes didn't last."[161] By 2002, other scientists agreed that the "flood" benefits had been only temporary. Jack Schmidt of Utah State University stated bluntly, "We're sort of back to where we were before the '96 flood. Some measures tell us we have less sand than before."[162] Others expressed concern that the sand that had built the temporary beaches had come from existing beaches rather than from the river bottom.[163] In addition, the experiment had been expensive. The flood cost at least $1.5 million in research on the effects and another $2.5 million in lost revenue from altered electricity generation.[164]

In spite of the disappointing results, the 1996 flood was seminal in at least one way. The flood did represent a willingness to try to repair past damages. One analyst said that the experimental flows "clearly signify the beginning of a new era in dam administration. The 1996 releases from Glen Canyon Dam represent the first time in history that a federal reclamation project was operated exclusively for the benefit of the environment."[165] MRC scientists tried another brief high flow in 1997 after heavy winter snows, again with little success. In fact, some said that the simulated floods may have actually reduced the size of beaches.[166]

An experiment in the summer of 2000 involved changes that to some extent were the opposite of those involved in a simulated flood. The logic for a low, steady, summer flow was to emulate the natural summer hydrograph and eliminate, temporarily, the impacts of daily fluctuations. The GCD operators released water at a steady 8,000 cfs between June and September of that year, followed by a four-day spike of 30,000 cfs in late September. As one would expect, the power company representatives were concerned about losing the ability to increase flows to meet peak demand during the summer months. Barry Gold told me later that the AMP participants therefore agreed to keep the experiment relatively underpublicized to avoid blame for any power outages.[167] The experiment had several effects. River water temperatures increased to about 67 degrees Fahrenheit almost immediately and stayed high through the summer months. That seemed to have a positive impact on vegetation. Scientists noticed some increases in vegetation in the summer but also a larger decrease with the fall spike. The impacts on sediment were minimal. In terms of recreational impacts, the lower flows translated to "fewer whitewater thrills" and several grounding incidents for rafts but also into more camping area on the beaches and warmer water for swimming in the canyon.[168] Overall, the results were mixed but not substantial enough to create momentum for permanent changes in summer operations.

In 2002, MRC scientists developed a new test flow plan that called for controlled floods during periods when sediment levels were high as well as high fluctuating flows in winter months to reduce trout spawning and thus mitigate the impact of trout on native fish species. Scientists fear that the non-native rainbow trout population that has thrived in the cold river water may be contributing to the decline of humpback chub by eating their young. The chub population declined from 8,300 in 1993 to just 2,000 in 2002.[169] Trout anglers, power authorities, and some others expressed wariness about the plan, but the AMWG recommended the plan to the secretary of the interior.[170] Once the plan was approved, the scientists moved to implement it the following January.

After the late summer monsoons brought new sand into the Colorado from its tributaries, particularly the Paria River located sixteen miles downstream of the GCD, the Bureau of Reclamation would lower the flows from the dam to keep the sediment from washing downstream. Then, dam managers would release flood-level flows for two days to lift the new sand onto beaches and then alternate between high and low flows each day for three months. It was hoped that the alternating flows would dry spawning grounds and kill trout eggs without harming the chub and other species that dwell more in side channels.[171] Managers dropped the two-day flood part of the experiment due to the fact that drought had reduced the available sediment, but they did use fluctuating flows beginning in January 2003. Unfortunately, the experiment did not slow the decline of the chub or other species.

Another experiment followed in 2004. Dam operators initiated the first sediment-triggered flood in November 2004 following three heavy rainstorms in September and October that they hoped had pushed nearly a million tons of sediment out of the Paria and other tributaries into the Colorado. GCD operators had kept releases low since September to try to maintain the new sand. This $3.5 million flood released four times the amount of water normally released at that time of year, with peak flows of about 41,000 cfs over a two-and-a-half-day period.[172] The USGS studies of the impact of the experiment showed mixed results for restoration. On the positive side, the flood showed some promise in moving sediment and rebuilding beach habitat. On the negative side, the river still eventually lost more sediment than it gained and the number of endangered fish in the Canyon declined dramatically after the flood.[173]

Back to the Drawing Board

The Glen Canyon Dam adaptive management plan was not operating in a vacuum. During the first decade of the twenty-first century, NPS personnel engaged in a range of restoration efforts at the Grand Canyon. They struggled to reduce air pollution from nearby coal-burning power plants as well as distant sources such as Los Angeles that had impaired air quality and reduced visibility from a natural average of about 143 miles to less than 100 miles in the 1970s.[174] The NPS continued to try to restrict aircraft fly-overs in order to reduce noise pollution in the canyon; however, those restrictions were fully controlled by the Federal Aviation Administration, and the limited progress made continued to be a source of some frustration to park officials.[175] The NPS has also attempted, at various times since the 1970s, to phase out motorized forms of travel, particularly from outboard motors on large rafts, that create more noise in the canyon. Those efforts have been stymied by political opposition, notably an amendment to the 1981 DOI appropriations bill by Orrin Hatch (R-Utah) banning the NPS from reducing commercial motorized trips.[176] Park officials also were frustrated in their efforts to drastically reduce automobile traffic to the South Rim, much of it to be replaced by a light-rail system. That idea was, according to one NPS official, "squashed pretty quickly."[177] Today, the NPS faces threats from uranium mining on land adjacent to the canyon, largely because of renewed interest in nuclear power.[178] As for better news, in a classic success story, the NPS restored California condors to the Grand Canyon area in the mid-1990s.[179] The condors have made a remarkable comeback and can now be seen most summer days soaring over the canyon walls. By 2008, more than fifty condors resided in northern Arizona.

NPS and FWS officials also were working on another restoration attempt specifically involving the Colorado River that stopped short of making fundamental changes to the Grand Canyon ecosystem. In late 2004, officials from the Department of the Interior and the state governments of Arizona, California, and Nevada announced a fifty-year, $620 million project to protect and restore endangered species living along 342 miles of the lower Colorado River, from Lake Mead to Mexico. The project grew out of the Lower Colorado River Multi-Species Conservation Program, an Endangered Species Act compliance program of the FWS, and ten years of negotiations with different groups of stakeholders. The goal of the program is to restore more than 8,100 acres of streamside thickets, wetlands, and backwater pools that provide crucial habitat for endangered species such as the southwestern willow flycatcher and the humpback chub.[180] By 2008, the program was achieving some positive results.[181] Still, in a comment reflective of the larger picture, Bennett Raley, DOI assistant secretary, admitted that the project stopped short of complete restoration. He added, "It's impossible to go back to a natural Colorado River, given the number of people who rely on it."[182]

Whatever the gains from the other projects, the attainment of real success on the Colorado River through the Glen Canyon Dam AMP remained problematic. In 2005, the USGS published a comprehensive report on the results of the AMP since the filing of the environmental impact statement. The fact that a decade had passed since inception provided what the agency called "an important opportunity to evaluate the effects of Glen Canyon Dam on resources of concern."[183] The scientists with the USGS were thorough in their assessment, citing some improvement in some areas, but their conclusions only fanned the fires of debate over reconsideration of possible options. The report stated, "Dam operations during the last 10 years under the preferred alternative of the MLFF (modified low fluctuating flow) have not restored fine-sediment resources or native fish populations in Grand Canyon, both of which are resources of significant importance to the program."[184]

Many groups involved with the AMP have grown increasingly dissatisfied. By 2008, the experiments had cost roughly $80 million,[185] and some local groups have criticized spending that kind of money on what they see as experiments to save a few species of fish. Others, particularly environmental groups, are distressed by the lack of lasting improvements. In 2002, some groups started hinting at legal action under the ESA for the first time since passage of the Grand Canyon Protection Act of 1992. Grand Canyon Trust director Geoff Barnard warned that if the AMP did not produce better results, "the alternative is to revert to a pitched battle."[186] After the disappointing outcome of the 2004 experiment, John Weisheit of the Living Rivers group said bluntly, "It makes sense to get rid of Glen Canyon Dam, because it's the one changing the ecosystem of the Grand Canyon. It's dying."[187] After the 2005 USGS report, advocates for dam removal heralded the document as evidence that other options to dam operations had to be reconsidered.

In 2006, some groups acted on the threat of litigation. The Center for Biological Diversity, the Arizona Wildlife Federation, Living Rivers, the Grand Canyon chapter of the Sierra Club, and Glen Canyon Institute filed suit in federal court against the Bureau of Reclamation on behalf of the endangered humpback chub. They argued that the bureau had violated and was continuing to violate the Grand Canyon Protection Act as well as the Endangered Species Act. In an interview with me in early 2007, Weisheit expressed disappointment with the bureau's handling of the adaptive management program. "All they're doing is protecting hydropower by majority vote," he said. "They've done nothing to help the resource; all they've done is waste hundreds of millions of dollars."[188] Similarly, Michelle Harrington, director of the rivers program for the Center for Biological Diversity, remarked that "the program wasn't working, so we decided we have to go back to the drawing board."[189]

The Bureau of Reclamation agreed to restudy the management program if the groups dropped their lawsuit, and a settlement was reached in September 2006. In November, the agency announced that it would develop a new EIS on the implementation of a long-term experimental plan for the Glen Canyon Dam and the Colorado River.[190] Both Weisheit and Harrington stated that their groups were pushing for consideration of the alternative that had not been fully addressed in the 1995 EIS: dam removal. In response to my question regarding whether that option would be considered, Harrington responded, "We're asking for it to be on the table, but we don't know if it will be."[191] Indeed, the environmental groups were urging their supporters to lobby Congress to end the ban on study of decommissioning of the dam.

Future debate over dam removal will inevitably include discussion of other issues, although the impact of those issues on the debate is not yet clear. One issue is water supply. After six years of drought, arguably due to climate change, the water level in Lake Powell had dropped to a record low of 3,555 feet in 2005, or about one-third of capacity.[192] In response, the seven states sharing the Colorado River water agreed to a new pact in late 2007 that amends the 1922 compact. The agreement calls for reductions in allocations for downriver states from Lakes Mead and Powell if the Department of the Interior declares a shortage of water in the river. State officials also pledged to negotiate in future water crises before pursuing litigation. While state and federal officials hailed the agreement, some environmental groups warned that it did not emphasize conservation nearly enough and only postponed dealing with the long-term problem.[193] A couple of years of good snowfall in the Rockies brought the lake level up somewhat by 2008, but global warming could renew the drought. In 2008, two scientists with the Scripps Institution of Oceanography released a study showing that Lake Mead had a 50 percent chance of becoming unusable by 2021.[194] Climate change, therefore, may affect water supply and allocation. The tiny portion of water that now makes it to Mexico as a polluted, intermittent stream could be even more imperiled by severe drought.[195]

A second issue potentially affected by climate change involves fish species. As mentioned earlier, the release of cold water from the Glen Canyon Dam has been detrimental to the endangered chub. Scientists have considered installing temperature control devices to provide warmer water for the chub, but such devices would be incredibly expensive. One silver lining of climate change may be higher temperatures of the water in a partially filled Lake Powell; some scientists estimate that water temperatures are already a couple of degrees warmer than they were a decade ago.[196] The warmer water could actually assist the chub, but it might also bring more warm-water predators. Either of those outcomes could affect arguments for and against draining Lake Powell.

A third issue involving climate change recalls the mixed image of dam removal described earlier. Some scientists suggest that with the increased risk of drought, most Americans will be more reluctant to dismantle large dams that at least suggest that water is being stored.[197] However, the disappearance of Lake Powell could remove one of the arguments against removing the GCD. A severe drought could act as the kind of focusing event described in chapter 1, altering the framing of the issue of dam decommissioning. Further, as one scientist told me, if climate change continues to prompt improvements in wind and solar power that make them economically competitive sources of energy, that could provide more ammunition for those seeking to decommission hydropower plants like Glen Canyon Dam.[198]

More about Process than Product

Not surprisingly, the most recent actions in the Glen Canyon Dam AMP did not involve dam removal but another experiment. At the AMWG meeting in late 2006, participants discussed another high-flow test. The discussion was prompted by the fact that, according to MRC scientists, two years of storms had deposited significant amounts of sediment in the tributaries; they did not, however, approve a high-flow experiment. Then, in August 2007, after a substantial monsoon season, the scientists suggested that the time was ripe for another simulated spring flood. The AMWG subsequently debated a proposal for a high-flow event in spring 2008 and then five years of steady flows between September 1 and October 31. Again reflecting the lack of any kind of cohesive coalition, only the NPS and the FWS and two environmental groups voted for the experiment. Nevertheless, the secretary of the interior's representative to the AMWG presented the idea to department officials, who approved it.

In early March 2008, the Department of the Interior conducted another simulated flood. Dam operators released high flows (peaking at 41,000 cfs) for three days through diversion tubes rather than through the generators. John Hamill, chief of the MRC, said that the purpose of the "flood" was "to reconsider whether or not this is a viable strategy for creating sandbars and habitats for native fishes."[199] On March 5 the Bureau of Reclamation launched the high flow with a major media event featuring Secretary of the Interior Dirk Kempthorne that was covered by national news outlets including NBC, NPR, and the *New York Times*. DOI officials promised that the high, beach-building flow would be followed by a steady flow in September.

I visited the Grand Canyon during the March event to see the "flood" and talk to some of those involved. Even from the South Rim, the river looked much larger and much muddier than usual. During the event, I talked to Jan Balsom, NPS deputy chief of science and resource management at the canyon and one of the planners for the 1995 environmental impact assessment. Just after the event and after he had returned from overseeing tests at river level, I spoke with Sam Spiller, the Lower Colorado River coordinator for the FWS. Both officials confirmed that the high flows had stirred up sediment in the tributaries and the main channel. As Spiller noted, this flood involved some kinds of sediment that might actually remain on restored beaches. Months later, other scientists expressed some optimism that the flows might have done some good by increasing sediment deposits.[200]

Nevertheless, many observers, including people involved with the AMP, were skeptical of the value of the test. A power company spokesman said the test would cost the company $4 million to purchase the lost power that it still needed to meet contracts.[201] From quite another perspective, NPS and FWS officials argue that what are needed are high flows on a nearly annual basis determined by the availability of sand for beach restoration rather than just every four or five years; moreover, they say, steady flows are needed in the summer months, not September. Steady summer flows would more closely resemble the natural hydrograph and would likely be more beneficial for the baby humpback chub. Steve Martin, NPS Grand Canyon superintendent, bemoaned the fact that he had been given only one day to submit comments on the 2008 experiment and questioned the overall intentions behind it: "It is not apparent where the 80 million dollars in research conducted over the last 10 years has been used in this decisionmaking process." He suggested instead that the timing of the experiment was driven by the hydropower companies, which did not want to lose any peak power in high-value summer months.[202] Martin expressed concern that the "flush" might do more harm than good to canyon resources such as endangered fish and archeological treasures.[203]

The timing does raise questions. When I asked Balsom about the timing of the steady flows, she answered, "We really need it in the summertime, and all the biologists say we need it in the summer, but the bureau said they need it in September to allow for power interests in the summer." In other words, the bureau did not want to interfere with hydropower generation in the hot summer months when peaking operations facilitate large profits; it would rather allow them in the "shoulder months" of September and October when the demand for power is less. Balsom added, "Any water past the generators is profit not realized."[204] Spiller said, "We preferably would have low flows in June or July, but the reason we don't do them then is that costs money for the power companies."[205] The U.S. Geological Survey confirmed that floods were timed to avoid impairing power generation during summer months.[206]

So, why bother with the experiment? More specifically, why would DOI officials approve the experiment when it had not been approved by the AMWG and any benefits to the river were dubious? Jeff Ruch, executive director of Public Employees for Environmental Responsibility, said that the experiment was "nothing but a green wash to mask another betrayal of the Grand Canyon by its political custodians."[207] This group also suggested that the awarding of exclusive rights to NBC News to film the event implied that that the real reason behind it was to gain publicity. Other environmental groups also were extremely critical. Nikolai Lash, water program director for the Grand Canyon Trust, said, "They're trying to make it appear that they're doing something beneficial when they're just doing it for appearances."[208] Balsom confirmed that the event was a very public example of the secretary of the interior taking action on behalf of the canyon. As she said, "The science should be focused on management questions, questions that really answer the 'so what' of the science. We have invested a lot into the studies without changes to management."[209] Many participants are not confident of any commitment to pursue a regular flow regime such as one involving adjusted summer flows. Lash agreed on the need for regular high maintenance flows but said that when votes were taken at the Glen Canyon Dam AMP meetings, the hydropower interests always prevailed.[210] The Grand Canyon Trust has filed a lawsuit to enforce a more consistent flow pattern.

What is the future of the adaptive management program, particularly now that stakeholders are again resorting to litigation? The 2005 USGS review of the program after ten years cited some progress, notably that "a substantial body of knowledge now exists for the Colorado River ecosystem in Grand Canyon." But the reviewers were also quite realistic, and their wariness about the challenges facing the program recalls the theoretical importance of involving the larger public, which is so emphasized in the argument of this book. The review concludes: "The overarching question is 'What will society do with the knowledge now available?'"[211]

Can those seeking changes at the canyon develop an effective coalition to create the conditions needed to involve the larger public? Three veterans of the AMP are not overly optimistic. Spiller remains reluctant to criticize the AMP, but he admits that real progress has been slow, noting the difficulties of dealing with "so many concerns involved." He was positive about the fact that all the groups are still talking and that the FWS, often relegated to a lesser role in similar procedures, was involved in the decisionmaking process. When I asked him whether the AMP might be aborted, he answered, "I hope not, because at least now we have a seat at the table."[212] In a later interview, he added, "There's value to collaborative processes; we don't always have to reach consensus."[213] Balsom, who had helped craft the 1995 EIS, offered a starker assessment: "We had great hopes when we finished the EIS, but this is now more about process than product. They provide funds for a multitude of studies without ever addressing resource conditions in the park." She added, "It's become way too political, entrenched with stakeholders, and not focused on the common goal."[214] As head of the MRC for years, Barry Gold was as responsible as anyone for building the AMP. When he left in 2002, he had high hopes for the program. Today, his assessment is similar to Balsom's: "There were some intractable conflicts at the canyon," Gold said, noting the frustration of trying to change the political calculus. "We tried but we couldn't get people to agree on the future vision of that ecosystem."[215]

Whatever the future vision of the ecosystem, it is unlikely to include dam removal anytime soon. The entrenched political interests mentioned above are adamantly opposed to the idea, but even the FWS and the NPS remain publicly uninterested in this option, thereby dimming the chances for an effective pro-removal coalition. Spiller expressed little enthusiasm for the idea. He noted that the biggest obstacle was not the loss of hydropower, but the loss of water storage capacity. "Unless we can clearly show that we can deliver as much water without the dam," he said, "it won't happen."[216] Balsom told me, "Most of us fully acknowledge that we won't get rid of that dam in our lifetimes."[217] Thus, the AMP will continue to pursue experimental projects as substitutes for the real restoration that could come with decommissioning the primary structure that controls the canyon.

Counting on a Miracle

Amazingly enough, John and I found a rock just before we floated into the main channel of the Colorado just above Havasu Creek Rapid. The dark, wet boulder didn't protrude much above the river surface, but we both managed to grab it and hang on. Within a few minutes, one of the

In the heart of the Grand Canyon, a raft pauses before entering Horn Creek Rapid on the Colorado River just downstream of Phantom Ranch.

most experienced rafters on the trip maneuvered his boat out to where we could climb aboard. After we both settled in, he remarked, "Funny thing is that I've never seen that rock before." John looked at me with an expression on his face that combined joy and awe. He still thinks it was a miracle, and who knows?

Whether or not we experienced a miracle that day at Havasu Creek, a miracle is what will be necessary to achieve restoration of the Colorado River in the Grand Canyon. In spite of Teddy Roosevelt's warning, people did "mar" this wonderful place, and only dramatic changes will return it to a more natural condition. In his song "Countin' on a Miracle," Bruce Springsteen sings, "We've got no fairytale ending."[218] Nor is one in sight any time soon for the repair effort at the Grand Canyon.

Yosemite Falls, viewed from Glacier Point

6
REPAIRING DAMAGE FROM PAST POLICIES

In the end, democratic policymaking is to be judged not by the errors it makes, but the errors it corrects.

—Bryan Jones and Frank Baumgartner, 2002

You have to stand up.

—Mike Finley, Yellowstone superintendent, when asked why he was willing to take a controversial position on reintroducing wolves, 1996

Given that the United States is a country that was established through revolution, a quintessentially radical act, the fact that making substantial changes to government policy is so difficult seems somewhat ironic. Indeed, many proposals for change are met with derision if not open hostility. Although human beings typically become accustomed to the status quo and are slow to embrace change, human beings also make mistakes. Consequently, they're often stuck with policies that, while they may have made sense at some time, no longer do. And perhaps the ultimate test for democratic governments, as the opening quote from two prominent political scientists observes, is how they fix past mistakes. How do they do it? The quote from Mike Finley comes from a conversation that we had at the time that wolves were being reintroduced in Yellowstone, a policy change that incited intense feelings and strong criticism of his actions as park superintendent. I had asked him why he was willing to stick his neck out. His answer, while much more succinct, is consistent with the overall argument in this book. Those seeking change have to do what it takes to make it happen.

This chapter provides some summary answers to the question of how proponents of change do what it takes. It briefly reviews the case studies in the previous chapters, generalizes beyond the empirical context of those cases, reconsiders the broader theoretical arguments on policy change, and offers some final comments.

Reviewing the Case Studies

Although created to be maintained in perpetuity, the national parks are not insulated from change. Some of the changes that have occurred in the parks, even when intended by past policies, have significantly damaged the natural environment. In recent decades, some U.S. policymakers have attempted to repair the damage. This book examines efforts to effect policy change at the crown jewels of the national park system, using a simple theoretical framework to anticipate and describe those efforts.

Creating Conditions for Change

The framework builds on arguments in the policy literature that posit the importance of building coalitions and creating the conditions necessary to expand the sphere of conflict over a proposed change. Effective coalitions elicit contributions from different actors, including interest groups, academics, scientists, and agency officials. They then have to create and use conditions that garner the support of the broader public for the proposed change. To review those conditions in the context of the case studies, I use a series of tables containing just a few comments that recall the discussion in the studies.

Table 6-1 summarizes the differences in the four cases in terms of issue definition. The columns display the image of the issue as defined historically, the definition of the issue used by advocates to try to change policy, and the current image or contested images shaping outcomes today. The depiction of wolves at Yellowstone shows the sharpest trajectory of change, from hated predator in the early part of the twentieth century to crucial component of an incomplete ecosystem in the 1980s to symbol of the wilderness today—or, as NPS official Doug Smith calls them, the poster children for the environmental movement

Table 6-1. *Shifts in Issue Definition of Repair Efforts*

Issue	*Historic image*	*Catalytic image*	*Current image*
Wolves in Yellowstone	Despised predator	Ecosystem savior	Symbol of wilderness
Cars in Yosemite Valley	Means of easy access to park	Cause of smog and traffic	Improvement in air quality or locking up of wilderness
Water flows in Everglades	Part of a worthless swamp	Part of a dying ecosystem	Rescue mission or costly pork barrel
Glen Canyon Dam at Grand Canyon	Power generator	Shaper of river ecosystem	Ecosystem destroyer or adjustable tool

Source: Author's illustration.

in the twenty-first century. Those seeking changes at Yosemite have tried to shift the image of cars from that of an easy means of access to the park to that of a principal cause of smog and congestion in the valley. Policy participants today view that image and the effort to reduce traffic as either improving the quality of the park or locking up the wilderness for a few eco-zealots. Repair of the Everglades took on a crusade-like tone as the pro-change coalition vowed to save not a worthless swamp but a vital ecosystem from near death. The final definition of the issue, however, remains contested by those who see it as either a crucial rescue mission or an expensive series of pork barrel projects. Calls to decommission Glen Canyon Dam are similarly debated. One side views the dam as destroying the downstream ecosystem while the other sees it as something whose operations can be adjusted to mitigate damage while still providing power.

Table 6-2 summarizes the differences in the four cases in terms of economic arguments about proposed repairs, providing a broad view of their benefits and costs and bottom line if one exists. The figures for Yellowstone indicate that the economic impact of reintroduction is a net positive by a wide margin. However, in immediate material terms, making an economic case for eliminating cars from the Yosemite Valley or removing the Glen Canyon Dam from the Colorado River in the Grand Canyon is difficult. To be persuasive, such arguments would have to

Table 6-2. *Economics of Repair Efforts*

Repair effort	*Benefit*	*Cost*	*Net*
Reintroduce wolves to Yellowstone	Annual tourism increases of $23 million	Annual livestock and hunting loss of < $1 million	Benefits > costs
Reduce cars in Yosemite Valley	Increase in aesthetic and air quality in park	Economic losses in gateway communities	Short-term net loss in economic terms
Restore water flows in Everglades	Increased tourism; restored ecosystem; new jobs	Lost sugarcane cultivation; impacts on development; implementation cost	Not clear, but promised funds defuse need to determine result
Remove Glen Canyon Dam	Natural river and ecosystem; aesthetic values	Lost power; impacts on Lake Powell	Short-term net loss in economic terms

Source: Author's illustration.

include aesthetic considerations and a focus on the long-term results. The economic analyses at the Everglades are still not clear, but most participants in the policy process interpret them as showing a net gain based on the inclusion of federal dollars that would enlarge the pie for all the different stakeholders.

Table 6-3 summarizes the differences in the four cases in terms of the scientific evidence regarding proposed repairs. At Yellowstone, the scientific evidence supporting reintroduction was as strong as the public image and the economic argument. One question involved the potential mixing of "experimental" and "natural" wolves, but that was more a legal question than a scientific one. The wolves provided the final scientific evidence by restoring ecological balance to a damaged ecosystem. The science at Yosemite supported reducing traffic if for no reason other than to improve the air quality, but even that argument was made less persuasive by an inconclusive EPA analysis. At the Everglades, some scientists doubted the worth of the CERP, but decisionmakers were finally persuaded by the promise of an adaptive management regime that could

Table 6-3. *Scientific Evidence Regarding Repair Efforts*

Repair effort	*Arguments for proposal*	*Caveats*	*Final scientific take on proposal*
Reintroduce wolves to Yellowstone	Huge improvements in ecosystem	Experimental status	Thriving ecosystem and wolf population
Reduce cars in Yosemite Valley	Improved air quality	Insufficient information on extent of improvement	Unclear impacts
Restore water flows in Everglades	Improved ecosystem	Skepticism about ecological impacts	Adaptive management
Remove Glen Canyon Dam	Improved survival of species and cultural sites	Unclear results	Inconclusive science

Source: Author's illustration.

be used to remedy mistakes as the restoration effort proceeded. Adaptive management is also an important component of the current efforts to change the operations of Glen Canyon Dam, after the ultimate alternative, removing the dam, was deemed unworthy of fuller scientific consideration.

Table 6-4 summarizes the differences in the four cases in terms of the commitment of the NPS and other formal institutions that might be involved in formulation and implementation of policy changes. Policy change at Yellowstone benefited from an extremely cooperative relationship between the National Park Service and the U.S. Fish and Wildlife Service. In recent years, the two agencies have had to deal increasingly with neighboring state governments, but the issue at that point becomes the delisting, not the reintroducing, of wolves. Policy changes at Yosemite require interaction with neighboring communities, and relations among the various stakeholders remain contentious. The repair efforts at Everglades and the Grand Canyon involve state entities that often have been reluctant to support dramatic changes and other

Table 6-4. *Commitment from National Park Service and Other Institutional Actors*

Repair effort	*Other institutional actors*	*Attitude of other institutional actors*	*Relations between NPS and others*
Reintroduce wolves to Yellowstone	U.S. Fish and Wildlife Service	Enthusiastic	Cooperative
Reduce cars in Yosemite Valley	Gateway communities	Hostile	Contentious
Restore water flows in Everglades	U.S. Army Corps of Engineers; Florida government agencies	Mixed	Evolving but still wary
Remove Glen Canyon Dam	U.S. Bureau of Recreation; basin states	Skeptical	Suspicious

Source: Author's illustration.

federal agencies whose histories suggest at least tension with the NPS; interactions among them therefore are often tense and complicated.

Outcomes

Given the differences summarized in the preceding four tables, the outcomes of the repair efforts are quite logical. Only when those seeking change are able to create and sustain all four conditions conducive to engaging the larger public are they likely to achieve the political support necessary for a major revision of traditional policies. I therefore argue that positive issue definition, compelling economic arguments, convincing scientific evidence, and the commitment to change of implementing agency personnel are all necessary; none by itself constitutes a sufficient condition for major change. If all conditions are not present, changes are likely to proceed only incrementally and inconsistently.

The reintroduction of wolves into Yellowstone National Park in the 1990s constituted seminal, dramatic, non-incremental change. Political institutions endorsed and funded the project, which was ultimately legitimized through important court rulings. The support of political actors has therefore been consistently strong, and implementation of the project has occurred faster and more easily than anticipated. The wolves have thrived, and they are in Yellowstone to stay. This project has corrected a policy that damaged the premier U.S. wilderness ecosystem. The only question now is what happens to the wolves as their range expands outside park boundaries.

Efforts to reduce and ultimately eliminate automobiles from Yosemite Valley have proceeded only incrementally. Without strong political support, NPS personnel have pursued changes largely on their own initiative. Cumulatively, those efforts, assisted by major flooding in 1997, have reduced some of the development and some of the traffic in the Yosemite Valley, but many participants remain frustrated at the slow pace of change. The promise made in the 1980 general management plan to eliminate cars remains an elusive objective.

Efforts to restore fresh water flows in the Everglades ecosystem also have proceeded slowly. Despite the highly publicized promises of 2000, political support has been evident only occasionally. Meanwhile, the complexity of the changes required has led to proposals to tackle one problem at a time. Progress has been made, but only in fits and starts. Whether or not a major land acquisition in 2008 will initiate a new period of accelerated implementation remains to be seen, but the record of the past decade suggests that proponents should be realistic in their expectations.

Real restoration of the Colorado River in the Grand Canyon remains unlikely. The one policy change that could produce restoration, removal or breaching of the Glen Canyon Dam, has received little if any support from mainstream political actors. Policymakers have instead pursued experimental flow regimes as a substitute for restoring the natural hydrograph, experiments that thus far have produced only marginal repairs. Short of removing the dam, the best that restoration advocates can hope for is a more systematic flow regime based on scientific recommendations, but such a change is not likely to receive support from the hydropower industry and its allies anytime soon.

Generalizing beyond the Case Studies

The natural areas examined in these case studies are exceptional in many ways, so an obvious question to ask is whether the lessons that they teach are exceptional too. Can we generalize to other natural areas in the United States, to natural areas outside the United States, and to policy changes in issues other than natural resources?

A New Era for Parks?

Has the last decade ushered in a new era in the management of national parks? When Secretary of the Interior Bruce Babbitt launched the first simulated flood on the Colorado River in 1996 in an attempt to mimic seasonal flows in the Grand Canyon, he proclaimed "a new era for ecosystems, a new era for dam management, not only for the Colorado but for every river system and every watershed in the United States."[1] Babbitt's strong statement was representative of many pronouncements regarding the efforts described in these chapters. Do those efforts signal a new era for national parks and other ecosystems in the United States, as Babbitt promised? Several factors suggest that the secretary's enthusiasm should be tempered somewhat.

First, the limited success of the efforts reviewed in the previous chapters, including at the Grand Canyon, where Babbitt made his announcement, suggests, if nothing else, the need for patience. Only under ideal conditions, such as at Yellowstone, do seminal policy changes occur fairly rapidly. Under other conditions, proponents of change have to be patient and persistent in their efforts to overcome what often amounts to decades, sometimes centuries, of traditional behavior. Given that most decisions, including appropriations, are based on the short term, that remains problematic for parks. As Yosemite's Linda Dahl said, if they are having trouble coming up with money to clean the bathrooms, what will be left for the research necessary to pursue ecologically oriented policy changes?

Second, efforts to make changes in natural areas will inevitably be affected in coming years by climate change. Warming trends have affected all of the parks under study here, through impacts on vegetation in Yellowstone, droughts in Florida and the Colorado Plateau, and the increased threat of forest fires at Yosemite. The impacts on policy change efforts may be direct, as when droughts occur, but they certainly will be at least indirect as park managers and other personnel are forced to allocate resources to relatively new and unanticipated issues.

Third, parks and other natural areas face another recent challenge in the declining visitation totals shown in figure 1-1. The trend toward reduced, or at least flat, visitation in recent years was evident at all four of the parks—the crown jewels of the park system—as well as in the overall system (again, totals were divided by 100 to make for comparable visuals). Not only are the totals down, but many visitors come from other countries and therefore cannot become part of a political constituency that seeks to protect U.S. national parks. Change advocates may be able to argue that their proposals will produce more natural areas and thus become more attractive to more visitors in coming decades, and they will likely point to the increase in tourism to see the wolves at Yellowstone as evidence. On the other hand, some will argue that "going natural" may mean "going too far"—that removing dams and reducing automobile access are radical acts that would alter the traditional behavior patterns that fostered support for and visitation to national parks.[2] Those seeking such changes may counter that argument with the observation that national parks constitute about 84 million acres, less than 4 percent of the area of the United States, and ask why the nation cannot seek pristine conditions on at least those lands. The fact remains, however, that the parks depend on visitors.

While these trends may seem somewhat daunting, evidence exists even in the limited progress of the repair efforts undertaken so far of a new way of thinking about parks and wilderness areas. A few decades ago, the prospect of reintroducing wolves or eliminating cars or restoring natural water flows or removing dams was pretty much only a pipedream of eco-zealots like Ed Abbey or a source of provocative demands from outspoken advocates like David Brower. Today, those possibilities are the subject of serious discourse and systematic research. Awareness of problems and recognition of opportunities have increased substantially. And changes have in fact occurred. Wolves are back in Yellowstone. Cars are less conspicuous in Yosemite and some other parks and hardly noticeable in Zion.[3] Natural water flows are not yet visible at the Everglades, but they are evident just north of the park on the Kissimmee River. Finally, while the Glen Canyon Dam still stands, hundreds of other dams, though admittedly smaller and less consequential, have been removed from other rivers in the United States. Indeed, the next major removal may well occur in a national park if the dams on the Elwha River in Olympic National Park are taken out as planned.[4]

Finally, President Obama and others in his administration certainly espouse a willingness to pursue substantial changes to traditional behavior, and the new administration could provide a huge boost for restoration efforts in the parks. Obama's secretary of the interior, Ken Salazar, promised a substantial amount of money, approximately $750 million of the larger stimulus package, to go specifically to restoration and protection efforts in parks.[5] The funds would help many repair efforts, including those described in this book. For example, the administration promised $96 million for Everglades restoration, an important increase given the slow federal commitment of funds to that effort, described in chapter 4.[6] As these cases have shown, the commitment of political leaders is crucial to the outcome of change efforts. Time will tell whether that commitment will continue with respect to the national parks.

Models for Projects around the World?

At the signing ceremony for the Everglades restoration legislation, Governor Jeb Bush said, "This is a model—not just for our country, but for projects around the world."[7] Comments by participants in many of these repair efforts, including from others involved with the Everglades project, frequently posit the U.S. experience as an example for projects all over the globe. Another way to generalize beyond U.S. parks is to consider application of the theoretical arguments in this book to repair efforts at natural areas in other nations. While a full discussion is beyond the scope of this book, a recent policy change in Canada exemplifies the use of the theoretical framework in another national context.

In just the last decade, change proponents reversed a century of policies at Banff National Park in the Canadian Rockies. When the park was established in 1887, railroads and other economic interests encouraged tourism and growth. The town of Banff actually predates the park, and over time it continued to expand into the natural areas of the ecosystem. That was problematic in that the town is located in the warm valley bottoms, which constitute less than 5 percent of the park but provide crucial habitat for many species. Whenever the park's managing agency (now Parks Canada) tried to restrict growth, it was overruled by political authorities with ties to Banff and railroad interests.[8]

In the mid-1990s, proponents for change, including those components of coalitions cited in the cases in this book (interest groups, academics, journalists, and agency officials) succeeded in creating the conditions described in the theoretical framework as necessary to attract strong political support for change. National newspapers and magazines shifted the image of Banff from that of a pristine wonderland to that of an area suffering from severe overcrowding, traffic, and pollution.[9] Reinforcing the new image, the International Union for the Conservation of Nature warned in 1994 that the park could lose its World Heritage status. The scientific community encouraged restrictions on growth to protect habitat and species. A prominent task force of scientists and others made more than 500 recommendations in 1996 to restore natural conditions in the park, the most explicit being to restrict the growth of the town.[10] Economic arguments for change were strengthened when citizens of Banff, concerned that their own property values might otherwise suffer, began to support limits, even voting in a nonbinding plebiscite to restrict their own growth. Political authorities, particularly officials in Parks Canada and the Ministry of Canadian Heritage, increased their commitment to making changes at the park. In 1998, Minister of Canadian Heritage Sheila Copps officially imposed limits on growth and called on Parks Canada to implement the recommendations. The growth limits are now firmly in place.[11]

Other Issue Areas

While my own empirical interests obviously focus on national parks and public lands, I argue that the theoretical arguments presented in this book can be applied to many issues involving policy change, although I readily admit that any extension to other issue areas will necessarily be preliminary.

One possible application of the theoretical framework involves climate change policies by state governments. As Barry Rabe and other scholars have documented, much of the action in climate change policy in the United States is now occurring at the state level.[12] Beginning in the late 1980s and early 1990s, many state-level actors began changing their energy policies to address climate change concerns, through actions such as developing portfolio standards requiring that a certain percentage of electricity come from renewable sources by a certain date. The variance in state behavior is substantial.

I argue that the theoretical framework presented here can be useful in assessing that variance. Issue framing in states, a dimension used by Rabe in his study of the formulation of such policies, varies; sometimes action is portrayed as a response to an environmental threat, sometimes as a response to an economic threat.[13] The economic arguments vary from state to state as well, with some states having much more entrenched fossil fuel industries than others. The political support for changing energy policy will also be determined by the receptivity of different states to scientific evidence regarding climate change. Although too simplistic, a first attempt to differentiate states could separate traditionally red from traditionally blue states, as the latter usually are more receptive to scientific arguments about climate change than the former. How could we differentiate state-level changes in climate change policy in terms of agency commitment? As Rabe discusses, states with active climate change programs generally have one official perceived to be the authority on the efforts.[14] States without an identified authority will be more likely to experience the interagency complications that make implementation of policy changes more difficult.

It also is possible to use the theoretical framework to differentiate policy changes at the national level—for example, changes regarding use of nuclear power in the United States. Over the last four decades, the political support for nuclear power has shifted from strong to weak and now back to strong again. Each time, those shifts were driven by changes in issue definition (energy needs to safety concerns and now back to energy needs), economic arguments (changes in the affordability of nuclear power plants), and scientific evidence (based on technological innovations).[15] Such an argument would be more compelling if changes in an issue area such as nuclear energy was compared with less dramatic changes in another issue area, such as health care, but again such a contrast is beyond the scope of this work.

Extending the Theoretical Arguments

The theoretical arguments are at least somewhat versatile, but what does their application in this volume say about broader discussions in the literature and future efforts to change policies? I suggest extensions along the lines of the framework developed in chapter 1 and used in the case studies.

Increasing the Efficacy of Coalitions

The literature on interest groups and advocacy coalitions is, as discussed in chapter 1, thorough and rich. My ego is not so big nor is this study so thorough as to suggest any major revisions to it here. I do think that this volume shows, as suggested by the advocacy coalition framework, that, first, coalitions can increase their efficacy by being inclusive. When interest groups pursuing policy change are willing to engage or even recruit journalists, for example, the latter can help change the definition of an image that sustains the status quo. When groups solicit or even just cite the work of academics, their arguments gain credibility among the larger public. When they show a willingness to work with rather than just criticize or demonize agency officials, they are more likely to be able to find common goals.

Second, just as inclusiveness shows some flexibility on the part of change advocates, so too does a willingness to use "venue shopping" to pursue policy changes. Political scientists increasingly note the ability of groups to pursue alternative venues—for example the courts—when their actions prove inefficacious in other venues, such as state legislatures.[16] Again, different components of coalitions can prove helpful in pursuing alternative venues. For example, many restoration efforts, including some described in this book, have sooner or later ended up in court. Lawyers and legal scholars therefore can be quite useful actors in effective coalitions.

The Importance of Issue Definition

Some of the most compelling work in the policy change literature now focuses on the role of image or issue definition.[17] This study does not dispute its importance. How a potential policy change has been framed is shown to be a crucial factor in determining the strength of the public support for change in all of the cases examined here. Can the analyses in this project extend that line of research?

One major part of the framing literature grows out of experiments in the field of prospect theory, an area of research that examines how people behave when "the possible outcomes of a gamble can be framed either as gains or losses relative to the status quo."[18] The most succinct summation is that people behave differently when they think that they are gaining something than they do when they think that they are losing something: "Because losses loom larger than gains, the decision-maker will be biased in favor of retaining the status quo. . . . In general, loss aversion favors stability over change."[19] The temptation exists to try to apply these arguments to the simple logic of restoration. Restoration usually will involve either adding something that should be there (gaining something) or removing something that should not be there (losing something).

In terms of ecosystems, for instance, we can put wolves back into the Yellowstone ecosystem and we can take the Glen Canyon Dam out of the Grand Canyon ecosystem. According to prospect theory, change will be easier with the former and more difficult with the latter. So far so good. However, moving beyond such simple examples gets more challenging. One could say that policymakers are attempting to take cars out of Yosemite Valley and put fresh water flows back into the Everglades, but those examples are not quite as straightforward. Even if the National Park Service takes cars out of Yosemite, it will have to put buses (and possibly parking lots) in. At the Everglades, to put fresh water flows back into the system, some agricultural uses and some development will have to be taken out.

The fact that prospect theory does not accommodate the latter cases as neatly as it might does not suggest its lack of utility. Indeed, I would argue that the public support for the Everglades restoration plan was made much stronger by portraying it as putting something back into the ecosystem without much in the way of trade-offs—or, as some of the policymakers quoted in the Everglades discussion suggested, expanding the pie. That argument is consistent with studies of negotiations that generally show "that positively framed bargainers are more cooperative, more likely to settle, achieve greater profits, and achieve greater joint benefits than negatively framed bargainers."[20] At the same time, I realize that such applications are a bit too simple. Recent research has shown that impacts from framing also depend on other factors, such as uncertainty and risk.[21] In general, applying prospect theory to real world situations has been difficult.[22]

To summarize then, the case studies in this work, particularly at Yellowstone and the Everglades, show that issue definition has had an impact on public opinion and in mobilizing previously uninterested individuals. The framing of issues at Yosemite and the Grand Canyon has been much less clear and has been contested by counter images of the consequences of proposed policy changes. These examples not only provide empirical evidence supporting recent models of policy change, but they also are consistent with recent work in political science on the importance of framing for public opinion and policy choices.[23]

The Limitations of Benefit-Cost Analysis

Economic debates over proposed policy changes almost inevitably involve comparisons of likely benefits and costs under the status quo and under alternative policy regimes. Although the directives were not always followed, legislation such as the National Environmental Policy Act and executive orders under several presidents required the use of benefit-cost analyses in conducting environmental impact assessments and in other procedures.[24]

Benefit-cost analysis (BCA) provided an explicit and compelling argument for change at Yellowstone, although the arguments it provided in the other case studies in this book were less definitive. Proponents of change and even those who analyze proposals often find benefit-cost analysis frustrating. Miller's attempt, discussed in chapter 5, to produce an analysis of the proposal to decommission Glen Canyon Dam is an apt example.[25] Further, the history of the use of this tool is checkered, because it has often been misused to justify what are in fact economically unjustifiable policies.[26] However, while those in charge do not need such an analysis to protect their projects, BCA has often been used as an effective tool by "those not in control of the political process."[27] When David Brower and others fought the proposed dams in the Grand Canyon, they used economic analyses quite effectively. The discussion in this book thus joins a fairly heated debate over the limitations of the tool. Several arguments regarding the use of BCA are especially relevant in discussing policy changes such as restoration of natural ecosystems.

One issue is that benefits and costs involve future impacts. The use of BCA is much more straightforward with projects that have more immediate costs and benefits, and its use in discussing efforts to repair national parks, which have been created to benefit future generations, is inevitably controversial. To adjust the calculations for the long term, analysts have to use discount rates to calculate the net present value of benefits and costs in future years. However, as the historical record shows, those rates are easily manipulated to under- or overestimate future values.[28] Some techniques can improve the use of discount rates. For instance, those doing the benefit-cost analysis can use sensitivity analysis by adjusting the discount rate figure 10 to 20 percent to see whether they get the same outcome.[29]

A second issue is that efforts to restore natural ecosystems always involve nonmarket goods that are not easily priced. As if describing some of the promised benefits of the projects discussed in this work, prominent BCA critic Steven Kelman writes, "Peace and quiet, fresh-smelling air, swimmable rivers, spectacular vistas—are not traded on markets."[30] Economists have developed different methods to calculate values for such goods, including hedonic pricing (inferring impacts based on changes in the value of market goods such as property), travel costs (how much people pay to travel to affected places), and willingness-to-pay surveys.[31] Critics are skeptical of such tools, saying that too many other variables affect the first two and questioning the validity of the third, especially given typically low response rates and the fact that "talk is cheap" in surveys, which do not impose real costs on respondents.[32] To alleviate some of the skepticism and make a BCA as accurate as possible, all such analyses for a policy change proposal should be subjected to external peer review.

While economists have offered methods to help calculate the value of some nonmarket goods, a third and related issue is that, at least according to Kelman and others, some goods are "priceless" and have "infinite value."[33] To some, attempting to measure things like the value of life—or even the value of the knowledge that a natural river exists, as in the case of removing the Glen Canyon Dam—is not only impractical but unethical.[34] Nevertheless, economists and insurance companies are asked to do such calculations frequently, as in the recent case of compensating the families of victims of the 9/11 attacks. If "priceless" values are involved, some defenders of BCA suggest that they "be treated as additional considerations" but only if it is essential to do so.[35]

Certainly, benefit-cost analysis was a factor in determining the political support for the repair jobs described in this book, and it will continue to be used for such efforts in the future. Further, as one economist argues, BCA can be a useful tool for deciding between alternatives. For example, suppose that everyone agrees that Yosemite Valley is priceless; BCA can then be used to help differentiate between various ways of protecting that priceless resource.[36] Bureaucrats such as NPS officials will be more effective in pursing restoration efforts if they develop more expertise in pursuing their own benefit-cost analyses. Agencies need more resources to develop such capabilities and organizational cultures that are friendlier to economic tools such as benefit-cost analysis.[37] That is applicable not only to the NPS in the context of the case studies in this book but also to other agencies involved in policy change proposals.

Dealing with Scientific Uncertainty

Policy change necessarily involves unknown future conditions and outcomes. However, the scientific evidence regarding possible repairs at the four parks in this study (and elsewhere) often is too tentative to compel strong political support. In recent years, policymakers have increasingly mandated or relied on adaptive management (AM) to deal with future uncertainties. As discussed in the Everglades and Grand Canyon cases, AM is designed to be a system of controlled experiments, careful monitoring, and integrated assessment to increase the learning and knowledge necessary to manage large ecosystems despite the uncertainties.[38]

A major concern regarding the use of AM is whether it is implemented sincerely or more as a political crutch that promises the ability to address problems in the future in order to ensure support of a proposal today. The Everglades and Grand Canyon cases figure prominently in such discussions. Certainly, as reflected by statements of participants in the case studies, AM played a crucial role in formulation and approval of the CERP and the Glen Canyon Dam environmental impact statement. Yet scientists involved in those efforts and observers have questioned the degree to which AM has been systematically applied in both.[39] As an NRC panel concluded in 2004 regarding the Everglades, AM is "currently more of a concept than a fully executed management strategy."[40] Further, as one analyst stated in general terms regarding policy change, "Much good has surely been accomplished in the name of adaptive management. . . . But there seems little doubt or debate that the idea—while raising important issues—has yet to fulfill its promise in practice."[41]

Given that adaptive management will surely continue to play a role in future policy change efforts, how can it come closer to fulfilling its promise? The NRC panel assessing the use of AM in the Everglades, of which I was a part, made several recommendations that could be adapted for restoration efforts in other contexts. First, the agency with primary responsibility for implementation of policy changes should have an agency-wide center to coordinate AM efforts or at least to facilitate comparison of different programs. In the context of the case studies in this book, the NPS would benefit from some institutional means to compare AM experiences at Everglades, Grand Canyon, Yosemite, and other park units. Second, obviously Congress would need to provide resources for such a center and provide resources more consistently for each program. Several scientists complained to me and other members of the NRC panel that adaptive management requires a long-term commitment of resources but that funding almost always occurs on a short-term basis. Third, scientists cannot be expected to make management decisions. One comment that I heard frequently at the Grand Canyon was that many members of the managing groups came to rely on the scientists to provide direction for the program when in fact the science was supposed to provide the information needed to pursue (or not) a particular direction, which was determined by the managers. Finally, for adaptive management to work, all involved, including different agency personnel and stakeholders, have to be willing to learn from the scientific experiments. As one scholar stated, "Most institutions are not very good at learning, especially when such learning would entail significant revision of their own goals and operating procedures."[42]

Mitigating Complexity

As mentioned in chapter 1, one of the bedrock lessons of the public policy literature is that the more complex the issue, the more challenging the implementation.[43] The theoretical framework used in this study suggests that complexity alone does not explain outcomes, but it does not dispute the impact of complexity on implementation, particularly at the Everglades and Grand Canyon, where repair efforts have been bogged down by interagency disputes and complicated ecosystems. Those pursuing policy changes have tried at least a couple of approaches to mitigate the impact of high complexity.

One approach involves breaking down a highly complex system into more manageable parts. As the chapter on the Everglades described, this option has been recommended by the National Academy of Sciences and endorsed by high-level political officials within the Department of the Interior. While taking such an approach is tempting, others quoted in the chapter warned that it carries some risks. Their warnings are supported by compelling arguments in the political science literature. First, if you break down a policy like the CERP into components, you are resigned to an incremental pace of change. One of the earliest critics in the political science literature of scholars who encouraged taking an incremental approach to policy change warned that "if policy outputs are divisible, they can be changed at the margins."[44] In other words, if policies are divided, change will likely occur only at the margins. Second, as studies of synergy show, the sum of a process can be greater than its parts. As NPS scientist Johnson warned of proposals to break the CERP into parts, "the chunks are all tied together, and it's hard to break it up in pieces."[45] Third, as both Johnson and the Corps of Engineers's Appelbaum warned, once a process is broken into pieces, the priority assigned to different components can be determined by largely political interests. The political science literature supports that view, often characterizing disaggregated policies as subject to heavy political manipulation.[46] Therefore, a logical recommendation is that if large-scale policy changes, such as restoration efforts, are disaggregated to be more manageable, policymakers need to take steps, such as mandating specific schedules, to insulate sequencing as much as possible from political interference.

Another approach to mitigating complexity follows directly from the need for interagency cooperation. As illustrated by the Everglades and Grand Canyon cases, high complexity requires a high degree of interaction between different agencies with different mandates. The literature on implementation of policy changes warns that poor interaction can doom a project.[47] One way to reduce tensions in such interactions is to develop and maintain professional norms of respect. NPS planner Dahl, a veteran of restoration efforts at both Everglades and Yosemite, spoke highly of the different participants in the former effort: "They all have dueling models," she said, "but at least they're sharing ideas."[48]

A related but different way to reduce tension is to get participants out of the professional sphere once in a while so that they can get to know each other as people, rather than just as representatives of different agencies. One thing that Barry Gold and others involved with the adaptive management program at the Grand Canyon tried to do was to get the various stakeholders together for outings, particularly river trips, on which they could get to know each other as well as the resource.

Summarizing Lessons

The purpose of the theoretical framework is to facilitate understanding of outcomes of efforts to change policies. While it is not intended to be a prescription for creating policy change, some lessons are worth emphasizing. Whenever possible, enhance the positive image of a potential change. Demand innovative and realistic cost-benefit analyses. Get long-term commitments for adaptive management processes. If possible, increase interagency cooperation without increasing the role of

PHOTO BY LYNN LOWRY

Although the author is wearing a Grand Canyon t-shirt, he's standing in the Lamar Valley at Yellowstone—a sign not of confusion but of his affection for both places.

politics. These lessons may sound so obvious as to be unnecessary, but they often are ignored.

Final Comments

I went back to Yellowstone in the fall of 2008. During just a week in the area, I made three separate trips to the Lamar Valley on three consecutive days. On the first trip, on a cool and cloudy afternoon, my friend Gary Ferguson and I stood and watched a pack of wolves bedded down in the grass above the Lamar River, just above a small pool where an elk stood in the water, obviously having found a temporary refuge from the predators on the hillside. The scene was stark, brutal, even visceral. I could not help but feel the temptation to try to assist the elk somehow, perhaps by scattering the wolves so that the waterlogged animal would at least have a running start. But as I well knew and as the preceding pages have described, humans have too often intervened in natural processes, with problematic results. We moved on. On the second day, Gary and I had to drive out of the park, but rather than taking the shorter, easier drive and avoiding a pending late fall snowstorm, we took the long route through the Lamar and got there just as the flakes started to fall. The snowy vistas looked like the product of an artistic collaboration between Albert Bierstadt and Winslow Homer, with dark, ominous skies enveloping a landscape of varied colors and light. My wife and I came back the next day. The snow-laden cold front had moved on, leaving us with a gorgeous afternoon in the valley. The elk and wolves were nowhere to be seen, but bison roamed across the light-brown hills and the golden aspen leaves along the river banks fluttered in a slight breeze. We lingered into the evening and basked in a glorious sunset.

Just as Yellowstone was the birthplace of the idea of a national park, it was the birthplace of this project, years before, when I watched wolves battle grizzly bears over an elk carcass in the Lamar Valley. I realized then that the reason that I could witness such an unforgettable event was that a political decision had been made to return wolves to the park, and I wondered whether the wolf project signaled a new way of thinking about and managing parks and other natural areas. This project has tempered any such optimism; repairing the damage from past policies has been and will continue to be difficult. Nevertheless, this latest trip reminded me, again, that some places are worth the struggle. And today, when I go back to the Lamar Valley, I know that yes, we can repair paradise.

APPENDIX: PARTIAL LIST OF INTERVIEWEES

Yellowstone

Colette Daigle-Berg, Yellowstone ranger
Bob Ekey, Greater Yellowstone Coalition
Mike Finley, Yellowstone superintendent
Gary Ferguson, Montana author
Steve Gehman, Wild Things Unlimited
Beth Kaeding, NPS planner
Betsy Robinson, Wild Things Unlimited
John Sacklin, Yellowstone management assistant
Chuck Schwartz, USGS manager
Doug Smith, NPS Wolf Project leader
John Varley, Yellowstone chief of resources
Mike Yochim, NPS planner

Yosemite

Linda Dahl, NPS chief of planning
Mike Finley, NPS superintendent
Bob Hansen, Yosemite Fund president
Chip Jenkins, NPS strategic planner
Jerry Mitchell, NPS chief of cultural resources

Al Runte, author and Yosemite historian
Hank Snider, NPS chief of resources

Everglades

Stuart Appelbaum, U.S. Army Corps of Engineers deputy for restoration management
Nick Aumen, NPS scientist
Michael Grunwald, author and Everglades historian
Alcee Hastings, U.S. Representative (D-Florida)
Robert Johnson, NPS director of the South Florida Natural Resources Center
Mark Kraus, Audubon Society
Tom MacVicar, consultant to sugar growers
John Ogden, South Florida Water Management District

Grand Canyon

Jan Balsom, NPS deputy chief of science and resource management at Grand Canyon
Steve Magnussen, Bureau of Reclamation director of western water operations
Barry Gold, director of Glen Canyon Dam Monitoring and Research Center
Michelle Harrington, Center for Biological Diversity
Norm Henderson, NPS research coordinator at Glen Canyon National Recreation Area
Linda Jalbert, NPS official at Grand Canyon
Clayton Palmer, Western Area Power Administration
Sam Spiller, U.S. Fish and Wildlife Service
Crystal Thompson, Central Arizona Project
Dave Wegner, former Bureau of Reclamation scientist
John Weisheit, founder of Living Rivers

NOTES

Chapter One

Epigraphs: Barack Obama from his acceptance speech at the Democratic National Convention, Denver, Colorado, August 28, 2008. Woodrow Wilson quote found in Gerald F. Lieberman, *3,500 Good Quotes for Speakers* (New York: Doubleday, 1983), p. 49.

1. "McCain's Suspect Winds of Change," *Toronto Star,* September 6, 2008 (http://atlanticcallcentres.com/comment/article/491810).

2. Theodore Lowi, *The End of Liberalism* (New York: Norton, 1969); Grant McConnell, *Private Power and American Democracy* (New York: Houghton Mifflin, 1966).

3. Charles E. Lindblom, "The Science of Muddling Through," *Public Administration Review* 19 (1959), pp. 79–88; Aaron Wildavsky, *The Politics of the Budgetary Process* (Boston: Little, Brown and Company, 1964).

4. Bryan D. Jones and Frank R. Baumgartner, *The Politics of Attention* (University of Chicago Press, 2005); Paul R. Schulman, "Non-incremental Policy Making," *American Political Science Review* 69 (1975), pp. 1354–70.

5. Frank R. Baumgartner and Bryan D. Jones, *Agendas and Instability in American Politics* (University of Chicago Press, 1993); Christopher J. Bosso, *Pesticides and Politics* (University of Pittsburgh Press, 1987); A. Lee Fritschler, *Smoking and Politics,* 4th ed. (Englewood Cliffs, N.J.: Prentice-Hall, 1989); John W. Kingdon, *Agendas, Alternatives, and Public Policies* (New York: HarperCollins, 1984).

6. Daniel B. Botkin, *Our Natural History* (New York: Putnam Books, 1995), p. 33.

7. National Research Council, *Restoration of Aquatic Ecosystems* (Washington: National Academies Press, 1992), p. 171.

8. See "About SER" (www.ser.org).

9. Rocky Barker, *Scorched Earth* (Washington: Island Press, 2005); Bob R. O'Brien, *Our National Parks and the Search for Sustainability* (University of Texas Press, 1999); Alfred Runte, *National Parks: The American Experience* (University of Nebraska Press, 1979); Richard West Sellars, *Preserving Nature in the National Parks* (Yale University Press, 1997).

10. Jeanne Nienaber Clarke and Daniel C. McCool, *Staking out the Terrain* (State University of New York Press, 1996), p. 82.

11. Survey by Daniel McCool cited in Ronald A. Foresta, *America's National Parks and Their Keepers* (Washington: Resources for the Future, 1984), p. 104.

12. James Q. Wilson, *Bureaucracy* (New York: Basic Books, 1989), p. 64.

13. Michael Frome, *Regreening the National Parks* (University of Arizona Press, 1992), p. 11.

14. Sellars, *Preserving Nature in the National Parks,* p. 267.

15. Clarke and McCool, *Staking out the Terrain,* p. 86; William R. Lowry, *Preserving Public Lands for the Future* (Georgetown University Press, 1998), p. 65.

16. John Freemuth, *Islands under Siege: National Parks and the Politics of External Threats* (University of Kansas Press, 1991), p. 703; Lowry, *Preserving Public Lands for the Future,* p. 55.

17. U.S. National Park Service, *National Parks for the 21st Century* (Washington: 1992), p. 9.

18. U.S. National Park Service, *National Parks for the 21st Century,* p. 9; see also Freemuth, *Islands under Siege.*

19. Barker, *Scorched Earth,* p. 8.

20. U.S. National Park Service, *National Parks for the 21st Century,* p. 17.

21. Daniel Kahneman and Amos Tversky, "Choices, Values, and Frames," *American Psychologist* 39 (1984), p. 348.

22. Lowi, *The End of Liberalism*; McConnell, *Private Power and American Democracy*; E. E. Schattschneider, *The Semisovereign People* (New York: Holt, Rinehart and Winston, 1960)

23. David Truman, *The Government Process* (New York: Knopf, 1951).

24. For a recent summary, see Paul A. Sabatier and Christopher M. Weible, "The Advocacy Coalition Framework," in *Theories of the Policy Process,* edited by Paul A. Sabatier (Boulder, Colo.: Westview Press, 2007), pp. 189–220.

25. Paul A. Sabatier and Hank C. Jenkins-Smith, "The Advocacy Coalition Framework: An Assessment," in *Theories of the Policy Process* (Boulder, Colo.: Westview Press, 1999), p. 119.

26. Hugh Heclo, "Issue Networks and the Executive Establishment," in *The New American Political System,* edited by Anthony King (Washington: American Enterprise Institute, 1978), pp. 87–124; John W. Kingdon, *Agendas, Alternatives, and Public Policies* (New York: HarperCollins, 1984).

27. Sabatier and Weible, "The Advocacy Coalition Framework," p. 209.

28. Ibid., p. 196.

29. Schattschneider, *The Semisovereign People,* p. 2.

30. Baumgartner and Jones, *Agendas and Instability*; Jones and Baumgartner, *The Politics of Attention.*

31. Frank R. Baumgartner and Bryan D. Jones, "Positive and Negative Feedback in Politics," in *Policy Dynamics,* edited by Baumgarter and Jones (University of Chicago Press, 2002), p. 12.

32. Ibid., p. 21.

33. Andrew C. Mertha, *China's Water Warriors* (Cornell University Press, 2008); Sarah B. Pralle, "Venue Shopping, Political Strategy, and Policy Change," *Journal of Public Policy* 23, no. 3 (2003), pp. 233–60.

34. Robert M. Entman, "Framing: Toward Clarification of a Fractured Paradigm," *Journal of Communication* 43 (1993), pp. 51–58; Kahneman and Tversky, "Choices, Values, and Frames"; Sheldon Kamieniecki, *Corporate America and Environmental Policy* (Stanford University Press, 2006), p. 59.

35. Jones and Baumgartner *The Politics of Attention,* p. 85.

36. Thomas A. Birkland, *After Disaster* (Georgetown University Press, 1997); Jones and Baumgartner, *The Politics of Attention,* p. 124; Paul A. Sabatier and Hank C. Jenkins-Smith, *Policy Change and Learning: An Advocacy Approach.* (Boulder, Colo.: Westview Press, 1993), pp. 22–23.

37. Baumgartner and Jones, *Agendas and Instability,* p. 16.

38. Gary C. Bryner, "Congress and Clean Air Policy," in *Business and Environmental Policy,* edited by M. E. Kraft and S. Kamieniecki (MIT Press, 2007), p.146; K. W. Easter, N. Becker, and S. O. Archibald, "Benefit-Cost Analysis and Its Use in Regulatory Decisions," in *Better Environmental Decisions,* edited by K. Sexton and others (Washington: Island Press, 1999), pp. 157–76; Daniel A. Mazmanian and Jeanne Nienaber, *Can Organizations Change?* (Brookings, 1979).

39. Paula Garb and John M. Whiteley, "A Hydroelectric Power Complex on Both Sides of a War," in *Reflections on Water,* J. Blatter and H. Ingram, eds., (MIT Press, 2001), pp. 213–37.

40. Scott W. Allard, "Competitive Pressures and the Emergence of Mothers' Aid Programs in the United States," *Policy Studies Journal* 32 (2004), pp. 521–44; F. S. Berry and W. D. Berry, "Innovation and Diffusion Models in Policy Research," in *Theories of the Policy Process,* 2nd ed., edited by Paul A. Sabatier (Boulder, Colo.: Westview Press, 2007), pp. 223–60; William R. Lowry, "Policy Reversal and Changing Politics: State Governments and Dam

Removals," *State Politics and Policy Quarterly* 5, no. 4 (2005), pp. 394–419; Jack L. Walker, "The Diffusion of Innovations among the American States," *American Political Science Review* 63 (1969), pp. 880–99.

41. Marc Reisner, *Cadillac Desert* (New York: Penguin Books, 1987), p. 295.

42. Tanya Heikkala and Andrea K. Gerlak, "The Formation of Large-Scale Collaborative Resource Management Institutions," *Policy Studies Journal* 33, no. 4 (2005), pp. 583–612; Michael E. Kraft and Daniel A. Mazmanian, "Conclusions," in *Toward Sustainable Communities,* edited by Mazmanian and Kraft (MIT Press, 1999), pp. 285–311; Julia M. Wondollek and Steven L. Yaffee, *Making Collaboration Work* (Washington: Island Press, 2000).

43. Kamieniecki, *Corporate America and Environmental Policy,* p. 60.

44. John D. Huber and Charles R. Shipan, *Deliberate Discretion?* (Cambridge University Press, 2002); Matthew D. McCubbins and Thomas Schwartz, "Congressional Oversight Overlooked," *American Journal of Political Science* 28 (1984), pp. 165–79; Terry M. Moe, "Control and Feedback in Economic Regulation," *American Political Science Review* 79 (1985), pp. 1094–116.

45. For example, see work on the Federal Reserve by Irwin Morris, *Congress, the President, and the Federal Reserve: The Politics of Monetary Policy-Making* (University of Michigan Press, 1999).

46. Wilson, *Bureaucracy,* p. 91.

47. Clarke and McCool, *Staking out the Terrain,* pp. 8–9.

48. Arun Agrawal, "Small Is Beautiful, but Is Larger Better?" in *People and Forests: Communities, Institutions, and Governance,* edited by C. Gibson, M. McKean, and E. Ostrom (MIT Press, 2000), pp. 57–85; Elinor Ostrom, "Reformulating the Commons," in *Protecting the Commons,* edited by J. Burger and E. Ostrom (Washington: Island Press, 2001), pp. 17–41.

49. Jeffrey L. Pressman and Aaron Wildavsky, *Implementation* (University of California Press, 1973); Paul A. Sabatier and Daniel Mazmanian, "The Implementation of Public Policy: A Framework for Analysis," *Policy Studies Journal* 8 (1980), p. 538.

50. Eugene Bardach, *The Implementation Game* (MIT Press, 1977); Pressman and Wildavsky, *Implementation*; Terry M. Moe, "The New Economics of Organization," *American Journal of Political Science* 28, no. 4 (1984), pp. 739–77; Michael M. Ting, "A Strategic Theory of Bureaucratic Redundancy," *American Journal of Political Science* 47, no. 2 (2003), pp. 274–92.

51. Sabatier and Mazmanian, "The Implementation of Public Policy," p. 547.

52. Bryan D. Jones and Frank R. Baumgartner, "Punctuations, Ideas, and Public Policy," in *Policy Dynamics,* edited by Baumgarter and Jones, p. 297.

53. Baumgartner and Jones, *Agendas and Instability*; Sabatier and Jenkins-Smith, "The Advocacy Coalition Framework."

Chapter Two

Epigraphs: Roosevelt quoted in A. B. Hart and H. R. Ferleger, *Theodore Roosevelt Cyclopedia* (New York: Roosevelt Memorial Association, 1941), pp. 673–74. Yellowstone official Doug Smith interviewed by the author, June 21, 1999.

1. Douglas H. Chadwick, "Return of the Gray Wolf," *National Geographic* 5 (1998), pp. 72–99.

2. Aubrey L. Haines, *The Yellowstone Story* (Yellowstone National Park: Yellowstone Library and Museum Association, 1977), p. xix.

3. Paul Schullery and Lee Whittlesey, "The Documentary Record of Wolves and Related Wildlife Species in the Yellowstone National Park Area prior to 1882," in *The Yellowstone Wolf: A Guide and Sourcebook,* edited by Paul Schullery (Worland, Wyo.: High Plains Publishing Company, 1996), p. 54.

4. Ibid., p. 43.

5. Douglas W. Smith and Gary Ferguson, *Decade of the Wolf* (Guilford, Conn.: Lyons Press, 2005), p. 7.

6. Chadwick, "Return of the Gray Wolf," p. 79.

7. L. David Mech, "Returning the Wolf to Yellowstone," in *The Greater Yellowstone Ecosystem,* edited by Robert B. Keiter and Mark S. Boyce (Yale University Press, 1991), p. 313.

8. Alston Chase, *Playing God in Yellowstone* (New York: Harcourt Brace Jovanovich, 1987), p. 121.

9. Ibid., p. 16; Schullery, *The Yellowstone Wolf,* pp. 68–71. For more on the creation of Yellowstone, see Judith L. Meyer, *The Spirit of Yellowstone* (Lanham, Md.: Rowman and Littlefield, 1996); Paul Schullery and Lee Whittlesey, *Myth and History in the Creation of Yellowstone National Park* (University of Nebraska Press, 2003).

10. Chase, *Playing God,* p. 121; Haines, *The Yellowstone Story,* p. 82.

11. U.S. Statutes at Large, vol. 17, chap. 24, pp. 32–33.

12. U.S. Code, Title 16, sec. 3.

13. Schullery, *The Yellowstone Wolf,* p. 77.

14. Chase, *Playing God,* p. 122.

15. Ibid.; Schullery, *The Yellowstone Wolf*; Smith and Ferguson, *Decade of the Wolf,* p. 7.

16. Chase, *Playing God,* p. 6.

17. Ibid., p. 28.

18. George M. Wright, Joseph S. Dixon, and Ben H. Thompson, *Fauna of the National Parks of the United States* (U.S. Government Printing Office, 1933).

19. Aldo Leopold, *A Sand County Almanac* (Oxford University Press, 1966), p. 130.

20. Schullery, *The Yellowstone Wolf,* p. 165.

21. Aldo S. Leopold, "Wildlife Management in the National Parks," report to Secretary of the Interior Udall, March 4, 1963, p. 32.

22. Rocky Barker, *Scorched Earth* (Washington: Island Press, 2005), p. 165.

23. U.S. Fish and Wildlife Service (U.S. FWS), *Final Environmental Impact Statement for the Reintroduction of Gray Wolves to Yellowstone National Park and Central Idaho* (Washington: 1994), appendix 1.

24. Smith and Ferguson, *Decade of the Wolf*, p. 29; J. L. Weaver, "The Wolves of Yellowstone," *Natural Resources Report* 14 (U.S. National Park Service, 1978).

25. Edward F. Bangs and Steven H. Fritts, "Reintroducing the Gray Wolf to Central Idaho and Yellowstone National Park," *Wildlife Society Bulletin* 24, no. 3 (1996), pp. 402–13; U.S. FWS, *Final Environmental Impact Statement*, appendix 1.

26. Endangered Species Act, sec. 4(b)(1)(A). "Commercial data" refers to trade data that provide information about specific species, such as a decline in the annual catch of a particular type of fish, or factors affecting species, such as habitat loss.

27. U.S. National Park Service, *Yellowstone Master Plan* (Department of the Interior, 1974), p. 25.

28. Chase, *Playing God*, pp. 130, 133.

29. Eugene Linden, "Search for the Wolf," *Time*, November 9, 1992, pp. 66–67.

30. U.S. National Park Service, *Statement for Management, Yellowstone National Park* (Department of the Interior, 1991), p. 28.

31. Interview with author, July 16, 1996.

32. Michael Barone and Grant Ujifusa, *The Almanac of American Politics 1990* (Washington: National Journal Groups, 1989), p. 1230.

33. John Skow, "The Brawl of the Wild," *Time*, November 6, 1989, pp. 14–15; U.S. FWS, *Final Environmental Impact Statement*, appendix 1. See also H.R. 2786, 1989, p. 2.

34. Barone and Ujifusa, *The Almanac of American Politics 1990*, p. 333.

35. Skow, "The Brawl of the Wild," p. 15.

36. S. 2674, 101st Cong., 2nd sess., May 22, 1990.

37. Young and Burns quoted in U.S. House of Representatives, *Restoration of Gray Wolves to Yellowstone National Park*, hearing before the Subcommittee on National Parks and Public Lands, 101st Cong., 1st sess., July 20, 1989, pp. 8 and 21.

38. Symms and Burns quoted in Sharon Begley, "Return of the Wolf," *Newsweek*, August 12, 1991, p. 49.

39. Gary Ferguson, *The Yellowstone Wolves* (Helena, Mont.: Falcon, 1996), p. 69.

40. Quoted in Skow, "The Brawl of the Wild," p. 14.

41. Barker, *Scorched Earth,* p. 177.

42. U.S. House, *Restoration of Gray Wolves,* p. 8.

43. Ibid., p. 23.

44. Todd Wilkinson, "Bringing Back the Pack," *National Parks* (May 1993), p. 27.

45. Interview with author, July 18, 1996.

46. Bangs and Fritts, "Reintroducing the Gray Wolf," p. 403; Wilkinson, "Bringing Back the Pack."

47. Reviewed in L. David Mech, "Returning the Wolf to Yellowstone," in *The Greater Yellowstone Ecosystem,* edited by Robert B. Keiter and Mark S. Boyce (Yale University Press, 1991), pp. 309–22.

48. Mech, "Returning the Wolf to Yellowstone," p. 311.

49. U.S. Fish and Wildlife Service, *The Northern Rocky Mountain Wolf Recovery Plan* (Washington: 1987).

50. U.S. House, *Restoration of Gray Wolves,* p. 42.

51. U.S. National Park Service, 1990. *State of the Rocky Mountain Region* (Washington: Department of the Interior, 1990), p. 9.

52. Skow, "The Brawl of the Wild," pp. 13–14.

53. Author's interview with Bob Ekey, July 18, 1996.

54. H.R. 2686, An Act Making Appropriations for the Department of Interior and Related Agencies, 102nd Cong., 1st sess., November 13, 1991.

55. S. 13230, U.S. Congress, *Congressional Record,* September 18, 1991.

56. Amendment 218, *Department of the Interior Fiscal Year 1991 Appropriations Bill* (GPO, 1991).

57. U.S. National Park Service, *Statement for Management, Yellowstone National Park,* p. 29.

58. Associated Press, "U.S. Could Return Wolf to Yellowstone," *New York Times,* May 5, 1994, p. A22; Wilkinson "Bringing Back the Pack," p. 27.

59. H.R. 2520, Department of Interior and Related Agencies Appropriations Act, 103rd Cong., 1st sess., November 11, 1993.

60. U.S. Fish and Wildlife Service, Environmental Impact Statement for the Reintroduction of Gray Wolves to Yellowstone National Park and Central Idaho (Washington: 1993), p. 6.

61. U.S. Fish and Wildlife Service, *Final Environmental Impact Statement,* appendix 1.

62. Endangered Species Act of 1973, sec.10(j)(2).

63. U.S. Fish and Wildlife Service, *Final Environmental Impact Statement,* abstract.

64. Bangs and Fritts, "Reintroducing the Gray Wolf," p. 403; U.S. Fish and Wildlife Service, *Final Environmental Impact Statement,* p. 4.

65. Bangs and Fritts, "Reintroducing the Gray Wolf," p. 403; U.S. FWS, *Final Environmental Impact Statement,* pp. 7–15.

66. Timothy Egan, "As Americans Adjust Nature, Wolves Get Pushed Around," *New York Times,* December 6, 1992, p. E5; Linden, "Search for the Wolf," p. 66; Tom Skeele, "Let Nature Fill the Niche," *High Country News,* May 17, 1993, p. 9.

67. Barry Meier, "Wolves Return to Montana, and the Greetings Are Mixed," *New York Times,* August 10, 1992, p. A1.

68. Interview with wolf project leader Doug Smith, June 21, 1999.

69. Skeele, "Let Nature Fill the Niche," p. 9.

70. U.S. FWS, *Final Environmental Impact Statement,* appendix 16, p. 65.

71. Interview with author, July 13, 1999.

72. L. David Mech, *The Wolf* (Garden City, N.Y.: Natural History Press, 1970), p. xvii.

73. Barry Lopez, *Of Wolves and Men* (New York: Scribner, 1978); Farley Mowat, *Never Cry Wolf* (Boston: Little, Brown and Company, 1963); Roger Peters, *Dance of the Wolves* (New York: McGraw–Hill, 1985).

74. Jim Johnson, *Wolves* (Minneapolis: New Rivers Press, 1993), p. 16.

75. Egan, "As Americans Adjust Nature," p. E5.

76. Quoted in Timothy Egan, "Ranchers Balk at U.S. Plans to Return Wolf to the West," *New York Times,* December 11, 1994, p. 1.

77. Begley, "Return of the Wolf," pp. 44–50; Michael Satchell, "The New Call of the Wild," *U.S. News and World Report,* October 29, 1990, p. 29; Skow, "The Brawl of the Wild," pp. 14–15.

78. Begley, "Return of the Wolf," p. 44.

79. Satchell, "The New Call of the Wild," p. 29.

80. Quoted in Begley, "Return of the Wolf," p. 46.

81. Quoted in Skow, "The Brawl of the Wild," p. 15.

82. Quoted in Sharon Begley, "Return of the Native," *Newsweek,* January 23, 1995, p. 53.

83. Quoted in Satchell, "The New Call of the Wild," p. 29.

84. Interview with author, June 21, 1999.

85. Egan, "Ranchers Balk at U.S. Plans," p. 44.

86. Ibid., p. 44.

87. Quoted in Egan, "Ranchers Balk at U.S. Plans."

88. Ibid., p. 44.

89. Begley, "Return of the Wolf," p. 50.

90. Egan, "Ranchers Balk at U.S. Plans," p. 1.

91. Begley, "Return of the Native," p. 53.

92. Interview with author, October 6, 2008.

93. James Salzman and Barton H. Thompson Jr., *Environmental Law and Policy* (New York: Foundation Press, 2003), p. 255.

94. For examples, see Judith A. Layzer, *The Environmental Case* (Washington: CQ Press, 2006), chapters 8, 16.

95. Michael Tennesen, "What to Do about Elk," *National Parks* 1 (1999), p. 24.

96. Ibid., p. 26.

97. Egan, "As Americans Adjust Nature," p. E5.

98. Eugene Linden, "Search for the Wolf," *Time,* November 9, 1992, pp. 66–67.

99. Begley, "Return of the Native," p. 53.

100. L. David Mech, "A New Era for Carnivore Conservation," *Wildlife Society Bulletin* 24, no. 3 (1996), p. 398.

101. Wilkinson, "Bringing Back the Pack," p. 25.

102. Begley, "Return of the Native," p. 53.

103. Ibid.

104. Satchell, "The New Call of the Wild," p. 29.

105. Ibid.

106. Egan, "Ranchers Balk at U.S. Plans," p. 44.

107. Skow, "The Brawl of the Wild," p. 13.

108. Mech, "A New Era for Carnivore Conservation," p. 398.

109. Quoted in Wilkinson, "Bringing Back the Pack," p. 26.

110. Timothy Egan, "Judge Allows Release of Wolves in West," *New York Times,* January 4, 1995, p. A12.

111. Wilkinson, "Bringing Back the Pack."

112. Quoted in Wilkinson, "Bringing Back the Pack," p. 29.

113. U.S. Fish and Wildlife Service, *Final Environmental Impact Statement,* appendix 9, p. 42.

114. Mark S. Boyce, "Ecological Process Management and Ungulates," *Wildlife Society Bulletin* 26, no. 3 (1998), p. 396.

115. William K. Stevens, "Wolf's Howl Heralds Change for Old Haunts," *New York Times* January 31, 1995, p. C1.

116. Timothy Clark and Anne-Marie Gillesberg, "Lessons from Wolf Restoration in Greater Yellowstone," in *Wolves and Human Communities,* edited by V. A. Sharpe, B. G. Norton, and S. Donnelle (Washington: Island Press, 2001), pp. 141, 144.

117. U.S. FWS, *Final Environmental Impact Statement,* pp. 4–19.

118. Ibid., pp. 4–17.

119. Ibid., appendix 5, p. 16.

120. Schullery, *The Yellowstone Wolf,* p. 212.

121. Quoted in Egan, "Ranchers Balk at U.S. Plans," p. 44.

122. Begley, "Return of the Native," p. 53.

123. Ibid.

124. Quoted in Begley, "Return of the Wolf," p. 48.

125. Associated Press, "U.S. Could Return Wolf to Yellowstone," *New York Times,* May 5, 1994, p. A22.

126. Interview with author, July 15, 1996.

127. Interview with author, June 21, 1999.

128. Interview with author, July 13, 1999.

129. Ronald A. Foresta, *America's National Parks and Their Keepers* (Washington: Resources for the Future, 1984); William R. Lowry, *The Capacity for Wonder* (Brookings, 1994).

130. Jeanne Nienaber Clarke and Daniel C. McCool, *Staking out the Terrain: Power and Performance among Natural Resource Agencies,* Suny Series in Environmental Politics and Policy (State University of New York, 1996), p. 122.

131. Ibid., p. 125.

132. Smith and Ferguson, *Decade of the Wolf.*

133. Interview with author, June 21, 1999.

134. B. P. Minn, "Attitudes toward Wolf Reintroductions in Rocky Mountain National Park," master's thesis, Colorado State University, 1977.

135. D. A. McNaught, "Park Visitor Attitudes toward Wolf Recovery in Yellowstone National Park," master's thesis, University of Montana, 1985.

136. A. J. Bath, "Identification and Documentation of Public Attitudes toward Wolf Reintroduction in Yellowstone National Park," in *The Yellowstone Wolf,* edited by Schullery, p. 198.

137. A. J. Bath, "Attitudes of Various Interest Groups in Wyoming toward Wolf Reintroduction in Yellowstone National Park," master's thesis, University of Wyoming, 1987.

138. Wilkinson, "Bringing Back the Pack," p. 26.

139. Satchell, "The New Call of the Wild," p. 29.

140. Quoted in Egan, "Judge Allows Release of Wolves," p. A12.

141. Interview with author, July 13, 1999.

142. Interview with author, June 21, 1999.

143. Associated Press, "U.S. Could Return Wolf to Yellowstone"; Egan, "Ranchers Balk at U.S. Plans."

144. Jim Robbins, "With Return of Wolves to West, Predatory Habits Bring Back Fear," *New York Times,* December 29, 1995, p. A22.

145. Quoted in Egan, "Judge Allows Release of Wolves," p. A12.

146. *New York Times* editors, "Eight Wolves Arrive in Wyoming as Battle over Them Goes On," *New York Times,* January 13, 1995, p. A 17.

147. Bangs and Fritts, "Reintroducing the Gray Wolf," p. 408; Gary Ferguson, *The Yellowstone Wolves* (Helena, Mont.: Falcon, 1996).

148. Ferguson, *The Yellowstone Wolves,* pp. 113–14.

149. Bangs and Fritts, "Reintroducing the Gray Wolf," p. 410; Ferguson, *The Yellowstone Wolves.*

150. Ferguson, *The Yellowstone Wolves*; Bob R. O'Brien, *Our National Parks and the Search for Sustainability* (University of Texas Press, 1999), p. 115; John Pickrell, "Wolves' Leftovers Are Yellowstone's Gain, Study Says,"

National Geographic News, December 4, 2003; personal communication with Doug Smith, October 21, 2008.

151. Quoted in William K. Stevens, "Triumph and Loss as Wolves Return to Yellowstone," *New York Times,* September 12, 1995, pp. C1, C4.

152. James Brooke, "Yellowstone Wolves Get an Ally in Tourist Trade," *New York Times* February 11, 1996, p. 20.

153. Interview with author, July 13, 1999.

154. Smith and Ferguson, *Decade of the Wolf*; William K. Stevens, "As the Wolf Turns: A Saga of Yellowstone," *New York Times,* July 1, 1997, p. C1.

155. Bangs and Fritts, "Reintroducing the Gray Wolf," p. 411.

156. Interview with author, July 15, 1996.

157. Brooke, "Yellowstone Wolves Get an Ally."

158. Katurah Mackay, "Court Ruling May Spell Doom for Park Wolves," *National Parks* (March-April 1998), p. 13.

159. Interview with author, July 15, 1996.

160. Mel Frost, "Wolves Are Here to Stay," *Greater Yellowstone Report* 17, no. 1 (2000), pp. 8–9.

161. U.S. National Park Service, press release, "YNP's Fiscal Year 1999 Budget," April 19, 1999, p. 1.

162. Interview with author, July 13, 1999.

163. Thomas McNamee, "The Wolf, Betrayed," *New York Times,* December 24, 1997, p. A17.

164. Quoted in Jim Robbins, "Ranchers Back Ruling on Wolves Out West," *New York Times,* December 14, 1997, p. 38.

165. *Wyoming Farm Bureau Federation et al.* v. *Bruce Babbitt,* Civil Case No. 94–CV–286–D, ruling of December 12, 1997.

166. Quoted in Margaret Kriz, "Wild about Wolves," *National Journal* 30, no. 5 (1998), p. 240.

167. Mackay, "Court Ruling May Spell Doom," p. 13.

168. McNamee, "The Wolf Betrayed," p. A17.

169. "Yellowstone Wolves and Diversity," *New York Times,* January 4, 1998, p. WK10.

170. Poll results found at www.forwolves.org/Ralph/maughan98.html.

171. Quoted in Kriz, "Wild about Wolves," p. 240.

172. Quoted in Associated Press, "Court to Keep Wolves in Rockies," *New York Times,* January 13, 2000, p. A1; see also ABC News, "Appeals Court: Wolves Can Stay in Yellowstone," January 13, 2000.

173. Statement of Secretary of the Interior Bruce Babbitt, quoted in "Court to Keep Wolves in Rockies," *New York Times,* January 13, 1999, p. A1.

174. Frost, "Wolves Are Here to Stay," p. 9.

175. Quoted in Thomas McNamee, *The Return of the Wolf to Yellowstone* (New York: Henry Holt, 1997).

176. William K. Stevens, "Wolves Are Establishing Strong Roots in Rockies," *New York Times* June 1, 1999, p. F3.

177. Interview with author, July 13, 1999.

178. Ralph Maughan, "More Howls in More Places," *Greater Yellowstone Report* 16, no. 1 (1999), pp. 14–15; Stevens, "Wolves Are Establishing Strong Roots."

179. Interview with author, August 29, 2005.

180. P. J. White and others, "Yellowstone after Wolves," *Yellowstone Science* 13, no. 1 (2005), pp. 34–41.

181. Smith and Ferguson, *Decade of the Wolf,* p. 193; White and others, "Yellowstone after Wolves," p. 35.

182. Kirk Johnson, "A Divide as Wolves Rebound in a Changing West," *New York Times,* January 2, 2008.

183. Douglas Gantenbein, "The Music of the Woods," *National Parks* 72, no. 1 (1998), pp. 26–29; M. Katherine Heinrich, "Wolf May Return to Olympic," *National Parks* 71, no. 7 (1997), p. 17.

184. Hope Hamashige, "Dog Virus May Be Killing Yellowstone Wolves," *National Geographic News,* January 17, 2006, p. 1.

185. White and others, "Yellowstone after Wolves," p. 35.

186. Clark and Gillesberg, "Lessons from Wolf Restoration," p. 135.

187. McNamee, *The Return of the Wolf*; Jim Robbins, "In Two Years, Wolves Have Reshaped Yellowstone Ecosystem," *New York Times,* December 30, 1997, p. F1; Smith and Ferguson, *Decade of the Wolf.*

188. Jim Robbins, "Hunting Habits of Yellowstone Wolves Change Ecological Balance in Park." *New York Times,* October 18, 2005, p. F3.

189. White and others, "Yellowstone after Wolves," p. 35.

190. Barker, *Scorched Earth,* p. 243.

191. D. Fortin and others, "Wolves Influence Elk Movements," *Ecology* 86, no. 5 (2005), pp. 1320–30.

192. Quoted in Robbins, "In Two Years, Wolves Have Reshaped Yellowstone Ecosystem," p. F6.

193. Interview with author, June 21, 1999.

194. Chris Conway, "Yellowstone's Wolves Save Its Aspens," *New York Times,* August 5, 2007; Scott Kirkwood, "Wolf and Consequences," *National Parks* 4 (2006), pp. 29–33; Pickrell, "Wolves' Leftovers Are Yellowstone's Gain."

195. Interview with author, July 9, 1999.

196. Todd Wilkinson, "Yellowstone's Bison War," *National Parks* (November 1997), pp. 30–33; George Wuerthner, "The Battle over Bison," *National Parks* (November 1995), pp. 37–40.

197. Todd Wilkinson, "Snowed Under," *National Parks* (January 1995), pp. 32–37; Elizabeth F. Daerr, "New Winter Use Plan at Yellowstone," *National*

Parks 73, no. 9–10 (September/October 1999), p. 11; Michael J. Yochim, "Snow Machines in the Gardens," *Montana: The Magazine of Western History* (Autumn 2003), pp. 2–14; Michael J. Yochim, *Yellowstone and the Snowmobile* (University Press of Kansas, 2009).

198. U.S. Government Accountability Office, *Yellowstone Bison: Interagency Plan* (Washington: 2008).

199. Susan G. Clark, *Ensuring Greater Yellowstone's Future* (Yale University Press, 2008).

200. Kirkwood, "Wolf and Consequences," p. 29.

201. Elizabeth G. Daerr, "A Howling Success," *National Parks* 11 (2000), p. 25.

202. J. W. Duffield, C. J. Neher, and D. A. Patterson, "Wolf Recovery in Yellowstone," *Yellowstone Science* 16, no. 1 (2008), p. 22.

203. Kirkwood, "Wolf and Consequences," p. 30; Scott Kirkwood, "Off the List," *National Parks* 82, no. 3 (summer 2008), pp. 12–13.

204. White and others, "Yellowstone after Wolves," p. 40.

205. Ibid., p. 37.

206. Kathy Etling, "Wolves Continue to Create Controversy," *St. Louis Post–Dispatch*, January 14, 2006, p. B27.

207. Duffield, Neher, and Patterson, "Wolf Recovery," p. 23.

208. Cat Lazaroff, "Helping Ranchers, Keeping Wolves," *Defenders* 83, no. 4 (2008), pp. 28–29.

209. Interview with author, October 6, 2008.

210. Heidi Ridgley, "Rocky Road ahead for Wolves?" *Defenders* (Spring 2008), p. 16.

211. White and others, "Yellowstone after Wolves," p. 37.

212. Johnson, "A Divide as Wolves Rebound."

213. Duffield, Neher, and Patterson, "Wolf Recovery," p. 25.

214. White and others, "Yellowstone after Wolves," p. 41.

215. Interview with author, August 29, 2005.

216. Interview with author, June 21, 1999.

217. Associated Press, "U.S. Loses Ruling on Gray Wolves," *New York Times*, February 1, 2005, p. A1; Kirk Johnson, "Limits Eased on Killing of Wolves," *New York Times*, January 4, 2005, p. A1.

218. Interview with author, August 29, 2005.

219. Pat Dawson, "Yellowstone Wolves: Embattled Again?" *Time*, October 10, 2007; Jim Robbins, "Resurgent Wolves Now Considered Pests by Some," *New York Times*, March 7, 2006, p. A11; Jim Robbins, "For Wolves, a Recovery May Not Be the Blessing It Seems," *New York Times*, February 6, 2007, p. C2.

220. Gary Ferguson, "The Big Bad Wolf," *Los Angeles Times*, April 30, 2008.

221. Jesse Harlan Alderman, "Idaho Gov. Wants to Kill All but 100 Wolves in State," *Summit Daily News,* January 12, 2007.

222. Dawson, "Yellowstone Wolves: Embattled Again?"

223. Johnson, "A Divide as Wolves Rebound."

224. Kirkwood, "Off the List," p. 12.

225. Quoted in Kirk Johnson, "U.S. Ends Protection for Wolves in Three States," *New York Times,* February 22, 2008.

226. Johnson, "U.S. Ends Protection for Wolves."

227. Patty Henetz, "Wolf's Death Stirs Fears for Species' Fate," *Salt Lake Tribune,* April 8, 2008.

228. Oren Dorell, "Wolf Kills in Rockies Cause Activists Alarm," *USA Today,* April 18, 2008, p. 3A; Ferguson, "The Big Bad Wolf."

229. "Wyoming Plan, Genetics Cause Wolf Relisting," *Casper Star Tribune,* July 21, 2008.

230. Michael Scott, "A New Path for Wolf Management," Statement for the Greater Yellowstone Coalition, August 4, 2008.

231. Interview with author, October 6, 2008.

232. *Defenders of Wildlife et al.* v. *U.S. FWS and U.S. DOI,* No. CV–08–56–M–DWM, U.S. District Court, May 2008.

233. Kirkwood, "Off the List," p. 13.

234. Joel Achenbach, "Administration Reopens Effort to De-List Endangered Gray Wolves," *Washington Post,* October 24, 2008.

235. Bob Dylan, *The Times They Are A-Changing* (New York: Columbia Records, 1964).

Chapter Three

Epigraphs: *Yosemite General Management Plan* (Washington: U.S. Department of the Interior, 1980), p. 3. Yosemite Chief of Planning Linda Dahl was interviewed by the author, July 17, 2008.

1. Quoted in William C. Everhart, *The National Park Service* (Boulder, Colo.: Westview Press, 1983), p. 113.

2. John Muir, *The Yosemite* (New York: The Century Company, 1912), pp. 7–9.

3. Everhart, *The National Park Service,* p. 7; George B. Hartzog Jr., *Battling for the National Parks* (Mount Kisco, N.Y.: Moyer Bell Limited, 1988), p. 5.

4. U.S. Congress 1864, 13 Statute, 325.

5. Alfred Runte, *Yosemite: The Embattled Wilderness* (University of Nebraska Press, 1990), p. 21; Joseph L. Sax, *Mountains without Handrails* (University of Michigan Press, 1980), p. 5.

6. Sax, *Mountains without Handrails,* p. 5.

7. Quoted in ibid., p. 21.

8. John Muir, *The Mountains of California* (New York: Century Company, 1894).

9. State of California, "Act of the Legislature of the State of California, approved March 3, 1905."

10. U.S. Congress 1890, 26 Statute, 650, Sec. 2.

11. U.S. National Park Service, *Alternative Transportation Modes* (Washington: Department of the Interior, 1994), p. iii.

12. Quoted in Richard H. Quin, "Highways in Harmony: The Story of Yosemite's Road System." Brochure prepared by the Historic American Engineering Record (Washington: Department of the Interior, 1991), p. 5.

13. James Bryce, "National Parks—The Need and the Future," in *University and Historical Addresses* (New York: Macmillan, 1913), as reprinted in David Harmon, ed., *Mirrors of America* (Boulder, Colo.: Roberts Rinehart, 1989), pp. 126–127, pp. 393–401; Viscount Bryce, *Memories of Travel* (New York: Macmillan Company, 1923), pp. 233–34; Everhart, *The National Park Service,* p. 65.

14. U.S. National Park Service, *Public Use of the National Parks* (Washington: 1968), table II.

15. Runte, *Yosemite: The Embattled Wilderness,* pp. 152–53.

16. Quin, "Highways in Harmony," p. 7.

17. Glynn Mapes, "Severe Overcrowding Brings Ills of the City to Scenic Yosemite," *Wall Street Journal,* June 24, 1966, p. 1.

18. Michael Frome, *Regreening the National Parks* (University of Arizona Press, 1992), p. 192; Sax, *Mountains without Handrails,* p. 73.

19. Runte, *Yosemite: The Embattled Wilderness,* p. 7.

20. Edward Abbey, *The Journey Home* (New York: Dutton, 1977), p. 144 (see back cover for *New York Times* quote).

21. Conrad L. Wirth, *Parks, Politics, and the People* (University of Oklahoma Press, 1980), p. 61.

22. Wirth, *Parks, Politics, and the People,* p. 278; Ronald A. Foresta, *America's National Parks and Their Keepers* (Washington: Resources for the Future, 1984), pp. 55, 96.

23. Hartzog, *Battling for the Parks,* p. 85.

24. Runte, *Yosemite: The Embattled Wilderness,* p. 203.

25. Quoted in Runte, *Yosemite: The Embattled Wilderness,* p. 204.

26. Quoted in Associated Press, "Plan to Cut Yosemite Congestion Is Called Lax by Conservationists," *New York Times,* December 1, 1979, p. 10.

27. *National Park Service Management of Concession Operations.* Hearings before the House Committee on Government Operations, 94th Cong., 1st sess. (Washington: Government Printing Office, 1975).

28. Frome, *Regreening the Parks,* p. 192.

29. U.S. NPS, *Yosemite General Management Plan.*

30. Philip Shabecoff, "Crowding and Decay Threaten U.S. Parks," *New York Times,* August 3, 1980, p. 1.

31. United Press International, "Park Service Planning Ban on Cars in Yosemite Valley," *New York Times,* November 1, 1980, p. 12.

32. U.S. NPS, *Yosemite General Management Plan,* p. 1.

33. U.S. General Accounting Office, *Issues Involved in the Sale of the Yosemite National Park Concessioner* (Washington: 1992), p. 6.

34. U.S. NPS, *Yosemite General Management Plan,* p. 3.

35. Ibid., p. 20.

36. Ibid., pp. 47, 31.

37. U.S. National Park Service, "Yosemite Valley/El Portal Comprehensive Design" (Washington: Department of the Interior, 1987), p. 4.

38. Interview with author, August 10, 1992.

39. *Hearings on Concessions Policy of the National Park Service,* Senate Subcommittee on Public Lands, National Parks, and Forests, 101st Cong., 2nd sess. (Washington: GPO, 1990), p. 115.

40. Philip Shabecoff, "Administration Seeks Greater Role for Entrepreneurs at Federal Parks," *New York Times,* March 29, 1981, p. 1; Bob R. O'Brien, "A Vision Sustained," *National Parks* (July-August 2002), p. 45.

41. Interview with author, August 27, 1996.

42. Quoted in Conservation Foundation, *National Parks for a New Generation: Visions, Realities, Prospects* (Washington: Conservation Foundation, 1985), p. 298.

43. Runte, *Yosemite: The Embattled Wilderness,* p. 217.

44. Quoted in Carl Nolte, "Yosemite Is 100 and Ailing," *San Francisco Chronicle,* September 28, 1990, pp. A1, A20.

45. U.S. NPS, "Yosemite Valley/El Portal Comprehensive Design," p. 1.

46. Ibid., p. 14.

47. Interview with author, August 27, 1996.

48. U.S. NPS, "Yosemite Valley/El Portal Comprehensive Design," p. 14.

49. Runte, *Yosemite: The Embattled Wilderness,* p. 215.

50. Interviews with Hank Snyder and Chip Jenkins, August 27, 1996.

51. Runte, *Yosemite: The Embattled Wilderness,* p. 222.

52. Interview with the author, July 28, 1992.

53. Interview with the author, August 27, 1996.

54. Frank R. Baumgartner and Bryan D. Jones, *Agendas and Instability in American Politics* (University of Chicago Press, 1993); Thomas A. Birkland, *After Disaster* (Georgetown University Press, 1997); Paul A. Sabatier and Hank C. Jenkins-Smith, *Policy Change and Learning: An Advocacy Approach* (Boulder, Colo.: Westview Press, 1993).

55. U.S. National Park Service, *Yosemite Valley Plan: Final Supplemental Environmental Impact Statement* (Washington: Department of the Interior, 2000), p. ES-4.

56. Interview with author, August 27, 1996.

57. U.S. National Park Service, *Alternative Transportation Modes* (Washington: Department of the Interior, 1994), pp. ES-2, ES-5.

58. U.S. NPS, *Yosemite Valley Plan,* p. 3-124.

59. Ibid., p. ES-10.

60. U.S. NPS, *Alternative Transportation Modes,* p. ES-5.

61. Ibid., pp. ES-6, ES-11.

62. Interview with Chip Jenkins, August 27, 1996.

63. U.S. NPS, *Alternative Transportation Modes,* p. ES-26.

64. Ibid., p. ES-16.

65. Quoted in Carl Nolte, "Few Favor Parking Lot in Yosemite Valley," *San Francisco Chronicle,* June 22, 1995, p. A21.

66. William R. Lowry, *Preserving Public Lands for the Future* (Georgetown University Press, 1998), p. 96.

67. U.S. National Park Service, *Draft Yosemite Valley Plan: Supplemental Environmental Impact Statement* (Washington: Department of the Interior, 2000), p. 10.

68. Wendy Mitman Clarke, "After the Flood," *National Parks* 3 (1999), p. 22.

69. Clarke, "After the Flood," p. 22.

70. For an interesting example, see Bob Madgic, *Shattered Air* (Short Hills, N.J.: Burford Books, 2005).

71. Clarke, "After the Flood," p. 25.

72. U.S. National Park Service, *Merced Wild and Scenic River Comprehensive Management Plan* (Washington: 2001), p. 11.

73. U.S. NPS, *Yosemite Valley Plan,* p. ES-1.

74. U.S. NPS, *Draft Yosemite Valley Plan.*

75. U.S. National Park Service, "Yosemite National Park Planning Update," vol. 16 (Yosemite: April 2000), p. 1.

76. U.S. NPS, *Yosemite Valley Plan,* p. ES-5.

77. Ibid., p. 2-44.

78. Ibid., pp. 2-73, 2-79, 2-80.

79. Ibid., pp. 2–247, 2-250.

80. David Brower, "David Brower on the Yosemite Valley Plan," *San Francisco Chronicle,* November 20, 2000.

81. U.S. NPS, *Yosemite Valley Plan,* pp. 2-254, 2-253.

82. U.S. National Park Service, *Merced Wild and Scenic River Draft Comprehensive Management Plan and Environmental Impact Statement* (Washington: 2000), p. ES-3.

83. U.S. NPS, *Merced Wild and Scenic River Comprehensive Management Plan,* p. 29.

84. U.S. NPS, *Yosemite Valley Plan,* p. ES-8.

85. Ibid., p. III-270.

86. David Louter, *Windshield Wilderness* (University of Washington Press, 2006), p. 169.

87. Clarke, "After the Flood."

88. U.S. NPS, *Yosemite Valley Plan,* p. III-360.

89. Brian Huse, quoted in John H. Cushman Jr., "Finding a Solution to Yosemite's Traffic Jams," *New York Times,* July 26, 1998, p. TR28.

90. Cushman, "Finding a Solution to Yosemite's Traffic Jams," p. TR28.

91. Associated Press, "Yosemite Park Plan Calls for Cutting Traffic by 60 Percent," *St. Louis Post-Dispatch,* March 28, 2000.

92. Evelyn Nieves and Matthew L. Wald, "Interior Department Plans to Reduce Traffic in Yosemite," *New York Times,* March 28, 2000, p. A14.

93. *Yosemite Valley Plan.* Oversight Hearing before the House Subcommittee on National Parks, Recreation, and Public Lands, 107th Cong., 1st sess., March 27, 2001 (Washington: GPO), p. 95.

94. Quoted in Ryan Dougherty, "Plan for Yosemite Valley Heating Up," *National Parks* 77, no. 7 (July-August 2003), p. 10.

95. U.S. NPS, *Yosemite Valley Plan,* p. 1-18.

96. Ibid., pp. 3-138–3-139.

97. Calculated from data in *Yosemite Valley Plan,* p. 3-159.

98. Ibid., p. 3-158.

99. Ibid., pp. 4.0-47.

100. Ibid., p. J-2.

101. Ibid., p. J-1.

102. Ibid., p. J-3.

103. Ibid., pp. J-3, J-5.

104. Ibid., pp. III-376, III-370, III-271.

105. Ibid., p. III-123.

106. Ibid., pp. 4.0-16, 3-65.

107. Ibid., p. 3-65.

108. Ibid., p. I-6.

109. Ibid., pp. 4.0-15.

110. Ibid., pp. III-123, III-125.

111. U.S. Environmental Protection Agency, Letter from Region IX commenting on the *Yosemite Valley Plan,* July 12, 2000.

112. The subsequent comments are found in the Comments Section (volume 3) of U.S. NPS, *Yosemite Valley Plan.*

113. U.S. NPS, *Yosemite Valley Plan,* p. III-371.

114. Ibid., p. YVPD-8900.

115. AScribe News, "National Park Service Releases Yosemite Valley Plan," press release of November 14, 2000, to environmental editors.

116. U.S. NPS, *Yosemite Valley Plan,* p. III-272.

117. Chuck Squatriglia, "Blueprint to Beautify, Restore Yosemite Tangled Up in Court," *San Francisco Chronicle,* January 21, 2007, p. A1.

118. Quoted in Associated Press, "Yosemite Park Plan Calls for Cutting Traffic by 60 pct.," *St. Louis Post-Dispatch,* March 28, 2000.

119. Jay Thomas Watson, "A Park for All (not just Homo Sapiens)," Letter to the editor in *New York Times,* January 25, 2002, p. A22.

120. Quoted in AScribe News, "National Park Service Releases Yosemite Valley Plan."

121. "Yosemite's New Look," *New York Times,* March 29, 2000, p. A24.

122. Squatriglia, "Blueprint to Beautify, Restore Yosemite Tangled Up in Court."

123. Brower, "David Brower on the Yosemite Plan."

124. U.S. NPS, *Yosemite Valley Plan,* p. ES-10.

125. Ibid., p. 2-79.

126. Greg Adair, "Yosemite under Siege," letter to the editor, *New York Times,* January 29, 2002, p. A20.

127. "Restoring Yosemite," *New York Times,* January 22, 2002, p. A18.

128. The subsequent comments are found in the Comments Section (volume 3) of U.S. NPS, *Yosemite Valley Plan.*

129. Ibid., p. YVPD-4297.

130. Calculated from data found at ibid., pp. III-497, III-498.

131. Ibid., p. III-491.

132. Ibid., p. III-493.

133. Michael Barone and Grant Ujifusa, *The Almanac of American Politics 2000* (Washington: National Journal Group, 1999), p. 183.

134. Letter dated June 30, 2000, included in U.S. NPS, *Yosemite Valley Plan,* volume 3.

135. Letter dated June 23, 2000, included in ibid.

136. Comment 20107, May 24, 2000, reprinted in ibid.

137. U.S. NPS, *Yosemite General Management Plan,* p. 1.

138. Interview with author, July 17, 2008.

139. Interview with author, August 27, 1996.

140. Quoted in AScribe News, "National Park Service Releases Yosemite Valley Plan."

141. O'Brien, "A Vision Sustained," p. 45.

142. U.S. National Park Service, "Yosemite National Park Planning Update," volume 20, June 2001 (Yosemite: 2001), p. 1.

143. *Yosemite Valley Plan.* Oversight Hearing before the House Subcommittee on National Parks, Recreation, and Public Lands, 107th Cong., 1st sess., March 27, 2001 (Washington: GPO), p. 2.

144. Ibid., pp. 8, 15, 33.

145. Ibid., pp. 50, 60, 64.

146. Ibid., pp. 49, 50.

147. Ibid., pp. 45–46.

148. Ibid., p. 16.

149. Ibid., p. 7.

150. Ibid., pp. 98, 113.

151. Ibid., pp. 102–03.

152. Quoted in Sarah Foster, "Protesters Rip Apart New Yosemite Plan," *WorldNetDaily,* May 13, 2003, p. 2.

153. Cited in Dougherty, "Plan for Yosemite Valley Heating Up," p. 11.

154. *Friends of Yosemite Valley* v. *Norton,* 348 F.3d 789, 803 9th Cir. 2003.

155. U.S. NPS, 2005 *Merced Wild and Scenic River Revised Comprehensive Management Plan,* p. ES-5.

156. Friends of Yosemite Valley, press release, December 10, 2007 (www.yosemitevalley.org).

157. Quoted in Foster, "Protesters Rip Apart New Yosemite Plan," p. 2.

158. U.S. National Park Service, "Yosemite National Park Planning Update," volume 26 (Yosemite: December 2003), p. 1.

159. Bruce Hamilton, "Introduction," in *Wild Yosemite,* edited by Susan M. Neider (New York: Skyhorse Publishing, 2007), p. xvii.

160. Interview with author, July 15, 2008.

161. Ibid.

162. Bonner Cohen, *The National Parks: Will They Survive for Future Generations?* Hearing before the House Committee on Government Reform, 109th Cong., 1st sess., April 22, 2005 (Washington: GPO), p. 119.

163. Interview with author, July 17, 2008.

164. Interview with author, July 15, 2008.

165. Interview with author, July 17, 2008.

166. Friends of Yosemite Valley, "The Truth about the Yosemite Valley Plan: A Development Plan, Not a Restoration Plan," 2004 (www.yosemitevalley.org).

167. Ibid.

168. *Friends of Yosemite Valley* v. *Kempthorne,* No. 07-15124, 9th Cir. 2008.

169. Interview with Chief Planner Linda Dahl, July 17, 2008.

170. Quoted in Squatriglia, "Blueprint to Beautify, Restore Yosemite Tangled Up in Court."

171. Interview with author, July 15, 2008.

172. Ibid.

173. "Out of the Wilderness," *The Economist,* July 12, 2008, p. 36.

174. Interview with author, July 17, 2008.

175. Ibid.

176. Interview with author, July 15, 2008.

177. "Cascades Dam Removal Lets River Flow Free in Yosemite," *Modesto Bee,* January 12, 2004.

178. Oliver R. W. Pergams and Patricia A. Zaradic, "Evidence for a Fundamental and Pervasive Shift away from Nature-Based Recreation," *Proceedings of the National Academy of Sciences of the USA* (Washington: National Academies Press, 2008); "Out of the Wilderness," *The Economist.*

179. "Out of the Wilderness," *The Economist,* p. 35.

180. William Cronon, "Foreword," in David Louter, *Windshield Wilderness* (University of Washington Press, 2006), p. xiii.

181. Interview with author, July 17, 2008.

182. Ibid.

183. Interview with Dahl, July 17, 2008.

184. See, for example, the song "Love Hurts," *The All-Time Greatest Hits of Roy Orbison* (Nashville: Monument Records, 1973).

Chapter Four

Epigraphs: Margery Stoneman Douglas and Randy Lee Loftis from their chapter in Douglas's book, *The Everglades: River of Grass* (Sarasota: Pineapple Press, 1988), p. 427. Stuart Appelbaum was interviewed by the author, January 10, 2008.

1. Francis N. Lovett, *National Parks* (Lanham, Md.: Rowman & Littlefield, 1998), p. 104; Alfred Runte, *National Parks: The American Experience,* 2nd ed. (University of Nebraska Press, 1987), pp. 108–109, 128–137.

2. Marjory Stoneman Douglas, *The Everglades: River of Grass* (Sarasota: Pineapple Press, 1947), pp. 5–6.

3. Norman Boucher, "Back to the Everglades," *Technology Review* (August 1995), p. 24. The other two are in Bulgaria and Tunisia.

4. James Henshall, *Camping and Cruising in Florida* (Cincinnati: Robert Clarke & Co., 1884; reprint, Port Salerno: Florida Classics Library, 1991), pp. 72–73.

5. Michael Frome, *National Park Guide* (Chicago: Rand-McNally, 1974), p. 27.

6. Jeanne Nienaber Clarke and Daniel C. McCool, *Staking out the Terrain* (State University of New York Press, 1996), pp. 17–49; Arthur Maass, *Muddy Waters* (Harvard University Press, 1951); Daniel A. Mazmanian and Jeanne Nienaber, *Can Organizations Change?* (Brookings, 1979).

7. Michael Grunwald, *The Swamp* (New York: Simon & Schuster, 2007), pp. 57, 128.

8. Fred Ward, "The Imperial Everglades," *National Geographic* 141, no. 1 (1972), p. 8.

9. Marjory Stoneman Douglas with Randy Lee Loftis, "Forty More Years of Crisis," in *The Everglades: River of Grass,* edited by Marjory Stoneman Douglas (Sarasota, Fla.: Pineapple Press, 1988, 391-427), p. 393.

10. S. S. Light, L. H. Gunderson, and C. S. Holling, "The Everglades," in *Barriers and Bridges to the Renewal of Ecosystems and Institutions* (New York: Columbia University Press, 1995), p. 129; Grunwald, *The Swamp,* p. 224.

11. Bill Gifford, "The Government's Too-Sweet Deal," *Washington Post,* January 9, 1994, p. C3.

12. Grunwald, *The Swamp,* p. 223.

13. Frome, *National Park Guide,* p. 28.

14. Boucher, "Back to the Everglades," p. 29; U.S. General Accounting Office, *Restoring the Everglades* (Washington: 1995).

15. Robert H. Boyle and Rose Mary Mechem, "There's Trouble in Paradise," *Sports Illustrated,* February 9, 1981, p. 88; John D. MacDonald, "Last Chance to Save the Everglades," *Life,* September 5, 1969, p. 63.

16. David McCally, *The Everglades: An Environmental History* (University Press of Florida, 1999), p. xviii.

17. U.S. Bureau of the Census, *Historical Statistics of the United States* (Washington: 1975), p. 26.

18. Light, Gunderson, and Holling, "The Everglades," p. 114.

19. Douglas and Loftis, "Forty More Years of Crisis," p. 392.

20. U.S. Bureau of the Census, *Historical Statistics of the United States,* p. 460.

21. Gifford, "The Government's Too-Sweet Deal," p. C3.

22. Light, Gunderson, and Holling, "The Everglades," p. 110.

23. Boucher, "Back to the Everglades," p. 29; Laura Helmuth, "Can This Swamp Be Saved?" *Science* News 155 (1999), pp. 252–53.

24. Phyllis McIntosh, "Reviving the Everglades." *National Parks* (January 2002), p. 30.

25. Grunwald, *The Swamp,* p. 241.

26. U.S. National Park Service, "Comprehensive Everglades Restoration Plan." Fact sheet (Homestead, Fla.: South Florida Natural Resources Center, 2007).

27. Light, Gunderson, and Holling, "The Everglades," p. 113; Loftis and Douglas, "Forty More Years of Crisis," p. 415.

28. Boucher, "Back to the Everglades," p. 25; Light, Gunderson, and Holling, "The Everglades," p. 113.

29. U.S. GAO, *Restoring the Everglades,* p. 5; U.S. National Park Service, "Restoration of Everglades National Park." Fact sheet (Homestead, Fla.: South Florida Natural Resources Center, 2007).

30. U.S. NPS, "Comprehensive Everglades Restoration Plan."

31. Ward, "The Imperial Everglades," p. 3.

32. William C. Everhart, *The National Park Service* (Boulder, Colo.: Westview Press, 1983), p. 65.

33. "The 50-Year War on the Everglades," *New York Times,* April 20, 1997, p. E14.

34. McCally, *The Everglades: An Environmental History,* p. 175.

35. Grunwald, *The Swamp,* p. 254.

36. MacDonald, "Last Chance to Save the Everglades," p. 59.

37. Light, Gunderson, and Holling, "The Everglades," p. 129.

38. Mark Derr, "Splendor in the Swamp," *Sierra (*July/August 1999), pp. 50–55; National Parks Conservation Association, "Recovery Proceeds for Everglades, Panthers," *National Parks* 69, no. 1 (January 1995), p. 15; Cyril T. Zaneski, "Anatomy of a Deal," *Audubon* 103, no. 4 (July 2001), p. 53.

39. Grunwald, *The Swamp,* p. 248.

40. David Helvarg, "Destruction to Reconstruction: Restoring the Everglades," *National Parks* 72 no. 3 (1998), p. 27; Light, Gunderson, and Holling, "The Everglades," p. 114.

41. Light, Gunderson, and Holling, "The Everglades," p. 127; U.S. NPS, "Restoration of Everglades National Park."

42. Boucher, "Back to the Everglades," p. 26; *Kissimmee River/Lake Okeechobee/Everglades Ecosystem,* hearing before the House Subcommittee on Water Resources, May 20, 1988, 100th Cong., 2nd sess. (Washington: Government Printing Office), pp. 9–14.

43. C. J. Walters, L. Gunderson, and C. S. Holling, "Experimental Policies for Water Management in the Everglades," *Ecological Applications* 2 (1992), pp. 189, 199.

44. Light, Gunderson, and Holling, "The Everglades," p. 115.

45. Mazmanian and Nienaber, *Can Organizations Change?* p. 24.

46. U.S. Army Corps of Engineers, *National Inventory of Dams,* 1995–96 ed. (Washington: 1996), p. 8.

47. Clarke and McCool, *Staking out the Terrain,* p. 42; Mazmanian and Nienaber, *Can Organizations Change?* pp. 44–46.

48. Department of the Interior, *Budget Justifications F.Y. 1991* (Washington: GPO, 1991).

49. Elizabeth Culotta, "Bringing Back the Everglades," *USA Today,* November 6, 1995, p. A1.

50. Glenn E. Haas and Timothy J. Wakefield, *National Parks and the American Public* (Washington: National Parks and Conservation Association, 1998), p. 31.

51. James Bovard, "The Great Sugar Shaft," *Freedom Daily* (Fairfax, Va.: Future of Freedom Foundation, 1998) (www.fff.org/freedom); Gifford, "The Government's Too-Sweet Deal," p. C3; Aaron Schwabach, "How Protectionism Is Destroying the Everglades," *Environmental Law Reporter* (December 2002).

52. W. Hodding Carter, *Stolen Water* (New York: Atria Books, 2004), p. 186; Gifford, "The Government's Too-Sweet Deal," p. C3; Schwabach, "How Protectionism Is Destroying the Everglades."

53. Bovard, "The Great Sugar Shaft."

54. Shireen I. Parsons, "Sour Deal: Federal Sugar Subsidy," *Charleston Gazette,* October 24, 1999.

55. Light, Gunderson, and Holling, "The Everglades," p. 113.
56. McCally, *The Everglades: An Environmental History,* p. 154.
57. U.S. Bureau of the Census, *Historical Statistics of the United States,* p. 461.
58. Grunwald, *The Swamp,* p. 281.
59. Ibid., p. 292.
60. Parsons, "Sour Deal."
61. Grunwald, *The Swamp,* pp. 308–09; Kim A. O'Connel, "Gore Unveils Everglades Plan," *National Parks* 70, no. 5 (1996), p. 13; Parsons, "Sour Deal"; Zaneski, "Anatomy of a Deal."
62. John H. Cushman Jr., "Noisy Fight over a Tax on Sugar," *New York Times,* November 2, 1996, p. 12.
63. McCally, *The Everglades: An Environmental History,* p. xvii.
64. Grunwald, *The Swamp,* p. 309.
65. Jeffrey Birnbaum and Russell Newell, "Fat and Happy in D.C.," *Fortune,* May 28, 2001, p. 94.
66. Mazmanian and Nienaber, *Can Organizations Change?* pp. 45, 57.
67. Grunwald, *The Swamp,* p. 293.
68. U.S. GAO, *Restoring the Everglades,* p. 3.
69. Douglas, *The Everglades: River of Grass,* p. 383.
70. MacDonald, "Last Chance to Save the Everglades," p. 61.
71. Grunwald, *The Swamp,* pp. 277–279.
72. Judith A. Layzer, *Natural Experiments* (MIT Press, 2008), p. 109.
73. Quoted in Robert H. Boyle and Rose Mary Mechem, "There's Trouble in Paradise," *Sports Illustrated,* February 9, 1981, p. 94.
74. Grunwald, *The Swamp,* p. 267.
75. Clarke and McCool, *Staking out the Terrain,* p. 45; U.S. House, *Kissimmee River,* pp. 1, 9.
76. U.S. House, *Kissimmee River,* p. 22.
77. Grunwald, *The Swamp,* p. 268.
78. Light, Gunderson, and Holling, "The Everglades," p. 143.
79. Ibid.
80. Grunwald, *The Swamp,* p. 357.
81. Interview with NPS scientist Robert Johnson, January 7, 2008.
82. Carter, *Stolen Water,* p. 89.
83. Dewitt John, *Civic Environmentalism* (Washington: CQ Press, 1994), chapter 5.
84. Grunwald, *The Swamp,* p. 303.
85. Culotta, "Bringing Back the Everglades."
86. Grunwald, *The Swamp,* pp. 292–302.
87. Derr, "Splendor in the Swamp," p. 51.
88. National Parks Conservation Association, "Lawsuit Filed to Save Sparrow," *National Parks* 74 no. 1 (January 2000), pp. 13–14.

89. National Parks Conservation Association, "Protecting Everglades National Park." Fact sheet posted online November 17, 1999 (www.npca.com); Laura Parker, "Big Plan Seeks Everglades Revival," *USA Today,* July 1, 1999, p. 3A.

90. Grunwald, *The Swamp,* p. 315.

91. Parker, "Big Plan Seeks Everglades Revival," p. 3A.

92. Grunwald, *The Swamp,* p. 310; Helvarg, "Destruction to Reconstruction," p. 26.

93. "Testing Time in the Everglades," *New York Times,* September 29, 1997, p. A18.

94. U.S. House, *Kissimmee River,* p. 6.

95. Light, Gunderson, and Holling, "The Everglades," p. 149.

96. Layzer, *Natural Experiments,* p. 113.

97. Ibid., p. 114.

98. Dawn Jennings, "South Florida Multi-Species Recovery Plan," *Endangered Species Bulletin* 23, no. 6 (1998), p. 22.

99. John H. Cushman Jr., "Clinton Backing Vast Effort to Restore Florida Swamps," *New York Times,* February 18, 1996, p. 1.

100. Quoted in Bill Lambrecht, "Everglades Is Winning Big in Campaign Battles," *St. Louis Post-Dispatch,* March 11, 1996, p. A6.

101. Ibid.

102. U.S. General Accounting Office, *South Florida Ecosystem Restoration: Task Force Needs to Improve Science Coordination to Increase the Likelihood for Success* (Washington: 2003), p. 17.

103. U.S. Army Corps of Engineers, "Draft Report Released" (Washington: 1998), p. 1.

104. Ibid.

105. John H. Cushman Jr., "U.S. Unveils Plan to Revamp South Florida Water Supply and Save Everglades," *New York Times,* October 14, 1998, p. 12.

106. John, *Civic Environmentalism,* p. 187.

107. Douglas, *The Everglades: River of Grass,* p. 385; MacDonald, "Last Chance to Save the Everglades," p. 58; Ward, "The Imperial Everglades."

108. Boyle and Mechem, "There's Trouble in Paradise," pp. 83-84.

109. Helvarg, "Destruction to Reconstruction," pp. 22–23.

110. "Congress's Duty in the Everglades," *New York Times,* May 12, 1997, p. A14. See also "The 50-Year War on the Everglades," *New York Times,* April 20, 1997, p. E14; "A Lifeline for the Everglades," *New York Times,* November 17, 1997, p. A22; "In Celebration of the Everglades," *New York Times,* December 26, 1997, p. A38.

111. Quoted in Grunwald, *The Swamp,* p. 334.

112. "Sugar's Sweet Deal," *New York Times,* April 27, 1997, p. E14.

113. Ibid. See also "Sugar's Latest Everglades Threat," *New York Times,* April 29, 1998, p. A24, and "Sugar's Sweetheart Deal," *New York Times,* October 30, 2007.

114. Bill Gifford, "The Government's Too-Sweet Deal," *Washington Post,* January 9, 1994, p. C3.

115. U.S. Bureau of the Census, *Statistical Abstract of the United States 2008* (Washington: 2008), pp. 535, 770.

116. Ibid., pp. 769, 770.

117. Statistics found at www.stateofflorida.com under "Florida Quick Facts," 2008.

118. John, *Civic Environmentalism,* p. 186.

119. Quoted in Layzer, *Natural Experiments,* p. 114.

120. Interview with author, January 7, 2008.

121. Ronald Keith Gaddie and James L. Regens, *Regulating Wetlands Protection* (State University of New York Press, 2000), p. 20.

122. Ibid.

123. Quoted in Grunwald, *The Swamp,* p. 320; see also Layzer, *Natural Experiments,* p. 117.

124. William K. Stevens, "Everglades Restoration Plan Does Too Little, Experts Say," *New York Times,* February 22, 1999, pp. A1, A15.

125. Laura Helmuth, "Can This Swamp Be Saved?" p. 252.

126. Quoted in Michael Grunwald, "An Environmental Reversal of Fortune," *Washington Post,* June 26, 2002, p. A01.

127. L. H. Gunderson, C. S. Holling, Stephen S. Light, *Barriers and Bridges to the Renewal of Ecosystems and Institutions* (Columbia University Press, 1995); Lance Gunderson, "Resilience, Flexibility, and Adaptive Management," *Conservation Ecology* 3, no. 1 (1999), pp. 1–11; Light, Gunderson, and Holling, "The Everglades"; C. J. Walters, L. Gunderson, and C. S. Holling, "Experimental Policies for Water Management in the Everglades," *Ecological Applications* 2 (1992), pp. 189–202.

128. Interview with author and other members of the NRC Panel on November 20, 2002.

129. John, *Civic Environmentalism,* p. 128.

130. Light, Gunderson, and Holling, "The Everglades," p. 149.

131. Clarke and McCool, *Staking out the Terrain,* p. 42.

132. Grunwald, *The Swamp,* p. 373.

133. Grunwald, *The Swamp,* pp. 369, 373.

134. *Everglades Restoration,* hearings before the Senate Committee on Environment and Public Works, January 7, May 11, and September 20, 2000, 106th Cong., 2nd sess., p. 25.

135. Matthew L. Wald, "White House to Present $7.8 Billion Plan for Everglades," *New York Times,* July 1, 1999, p. A14.

136. U.S. Army Corps of Engineers, "Draft Report Released," p. 1.
137. Interview with author, December 31, 2007.
138. U.S. NPS, "Comprehensive Everglades Restoration Plan."
139. National Research Council, *Adaptive Management for Water Resources Project Planning* (Washington: National Academies Press, 2004), p. 55.
140. Water Resources Development Act 2000, Title IV, Section 601(b), P. L. No. 106–541.
141. Grunwald, *The Swamp,* p. 317.
142. U.S. General Accounting Office, *Comprehensive Everglades Restoration Plan* (Washington: 2000), p. 6.
143. "Equity in the Everglades," *New York Times,* March 8, 1999, p. A16.
144. Martin Enserink, "Plan to Quench the Everglades' Thirst," *Science* 285, no. 9 (July 1999), p. 180.
145. Grunwald, *The Swamp,* p. 324.
146. Helmuth, "Can This Swamp Be Saved?" p. 256.
147. Layzer, *Natural Experiments,* p. 118.
148. "In Victory of Its Advocates, Everglades Restoration Is Moved Up," *New York Times,* April 8, 1999, p. A25.
149. U.S. GAO, *Comprehensive Everglades Restoration Plan,* p. 9.
150. National Research Council, *Adaptive Management,* p. 55.
151. Zaneski, "Anatomy of a Deal," p. 53.
152. U.S. Senate, *Everglades Restoration,* pp. 2, 6.
153. Ibid., p. 151.
154. Ibid.
155. Layzer, *Natural Experiments,* p. 118; Grunwald, *The Swamp,* pp. 327, 338–341; Zaneski, "Anatomy of a Deal," p. 51.
156. Mireya Navarro, "Florida City's Hopes for Airport Bog Down," *New York Times,* November 22, 1998, p. TR3.
157. Grunwald, *The Swamp,* p. 337.
158. Dana Canedy, "U.S. Bars Airport Near the Everglades," *New York Times,* January 17, 2001, p. A14.
159. Quoted in Grunwald, *The Swamp,* p. 3.
160. Lizette Alvarez, "Senate Approves $7.8 Billion Plan to Aid Everglades," *New York Times,* September 26, 2000, p. A1.
161. Ibid.; Lizette Alvarez, "House Approves Plan to Restore Everglades," *New York Times,* November 4, 2000, p. A11.
162. Layzer, *Natural Experiments,* p. 136.
163. Phyllis McIntosh, "Reviving the Everglades," *National Parks* (January 2002), pp. 30–34.
164. Interview with author and other members of the NRC Panel of November 20, 2002.
165. Quoted in McIntosh, "Reviving the Everglades," p. 34.

166. Jon R. Luoma, "Blueprint for the Future," *Audubon* 103, no. 4 (July 2001), pp. 66–69.

167. U.S. Government Accountability Office, *South Florida Ecosystem Restoration: Substantial Progress* (Washington: 2001), p. 3.

168. Michael Grunwald, "A Rescue Plan, Bold and Uncertain," *Washington Post,* June 23, 2002, p. A01.

169. Interview of November 21, 2002.

170. Tanya Heikkala and Andrea K. Gerlak, "The Formation of Large-Scale Collaborative Resource Management Institutions," *Policy Studies Journal* 33, no. 4 (2005), pp. 583–612; William R. Lowry, *Dam Politics* (Georgetown University Press, 2003), chapter 7.

171. Quoted in Grunwald, *The Swamp,* p. 313.

172. Ibid., p. 316.

173. Quoted in Frank Bruni, "Bush Carries Environment-Friendly Tone to Everglades," *New York Times,* June 5, 2001, p. A16.

174. "President Bush and the Everglades," *New York Times,* June 5, 2001, p. A22.

175. Blaine Harden, "National and State Politics Help Protect a Swamp," *New York Times,* April, 3, 2002, p. A1; see also "Mixed Grades on National Parks," *New York Times,* July 19, 2003, p. A12.

176. "Mixed Grades on National Parks," *New York Times,* July 19, 2003, p. A12; see also "Two Bushes and the Everglades," *New York Times,* November 10, 2004, p. A24.

177. Grunwald, *The Swamp,* p. 361.

178. Quoted in Abby Goodnough, "Effort to Save Everglades Falters as Funds Drop," *New York Times,* November 2, 2007, p. A2.

179. Interview with author, January 11, 2008.

180. Ibid.

181. Interview with author, November 20, 2002.

182. Interview with author, November 21, 2002.

183. Goodnough, "Effort to Save Everglades Falters," p. 2; Layzer, *Natural Experiments,* p. 124.

184. Interview with author, January 7, 2008.

185. Interview with author, January 10, 2008.

186. Comments at Everglades Coalition Conference on Captiva Island, January 11, 2008.

187. "Decision Time on the Everglades," *New York Times,* September 23, 2002, p. A24; see also Grunwald, *The Swamp,* p. 360.

188. Ryan Dougherty, "Sugar Deal Sours 'Glades Restoration," *National Parks* 77 no. 9 (September/October 2003), pp. 10–11.

189. "Two Bushes and the Everglades," *New York Times,* November 10, 2004.

190. Curtis Morgan, "Dazzling Wildlife Masks Ugly Picture," *St. Louis Post-Dispatch,* December 9, 2007, pp. B1, B5.

191. Interview with author, January 7, 2008.

192. Interview with author, November 21, 2002.

193. Ibid.

194. Interview with author, January 10, 2008.

195. Grunwald, *The Swamp,* p. 360.

196. Interview with author, January 7, 2008.

197. Layzer, *Natural Experiments,* p. 127.

198. Interview with author, July 17, 2008.

199. National Research Council, *Adaptive Monitoring and Assessment for the Comprehensive Everglades Restoration Plan* (Washington: National Academies Press, 2003), p. 9.

200. National Research Council, *Science and the Greater Everglades Ecosystem Restoration* (Washington: National Academies Press, 2003), p. 1.

201. National Research Council, *Adaptive Management,* p. 58.

202. Kim Todd, "Who Gets the Water?" *National Parks* 77, no. 1 (2003), p. 10.

203. Grunwald, *The Swamp,* p. 360.

204. National Research Council, *Aquifer Storage and Recovery in the Comprehensive Everglades Restoration Plan* (Washington: National Academies Press, 2001), p. 4.

205. National Research Council, *Regional Issues in Aquifer Storage and Recovery for Everglades Restoration* (Washington: National Academies Press, 2002), p. 3.

206. National Research Council, *Florida Bay Research Programs and Their Relation to the Comprehensive Everglades Restoration Plan* (Washington: National Academies Press, 2002), p. 3.

207. National Research Council, *Does Water Flow Influence Everglades Landscape Patterns?* (Washington: National Academies Press, 2003), p. 3.

208. National Research Council, *Florida Bay Research Programs,* pp. 26–27.

209. Grunwald, "A Rescue Plan," p. A01.

210. Grunwald, "An Environmental Reversal," p. A01.

211. W. Hodding Carter, "A Wetland Dying of Thirst," *New York Times,* July 15, 2004, p. A23.

212 Carter, *Stolen Water,* p. 255.

213. U.S. Environmental Protection Agency, *Everglades Ecosystem Assessment: Water Management and Quality* (Athens, Ga.: U.S. EPA Region 4 Science Division, 2007), p. 2.

214. Comments at Everglades Coalition Conference on Captiva Island, January 11, 2008.

215. U.S. Government Accountability Office, *South Florida Ecosystem: Some Restoration Progress* (Washington: 2007).

216. National Research Council, *Progress toward Restoring the Everglades* (Washington: National Academies Press, 2007), p. 12.

217. Ibid., p. 2; U.S. GAO, *South Florida Ecosystem: Some Restoration Progress,* p. 1.

218. Brian Skoloff, "Everglades Restoration Has Almost Stopped," Associated Press, November 24, 2007 (www.ocala.com/apps/pbcs.d11/article?).

219. Interview with author, December 31, 2007.

220. Comments at Everglades Coalition Conference on Captiva Island, January 11, 2008.

221. Interview with author, January 11, 2008.

222. "Hope for the Everglades," *New York Times,* November 10, 2007.

223. Carter, *Stolen Water,* p. 135.

224. Interview with author, January 7, 2008.

225. Interview with author, January 11, 2008.

226. Comments at Everglades Coalition Conference on Captiva Island, January 11, 2008.

227. National Research Council, *Progress toward Restoring the Everglades,* p. 59.

228. Grunwald, *The Swamp,* p. 363.

229. Grunwald, *The Swamp,* p. 366.

230. Interview with author, January 7, 2008.

231. "Sugar's Sweetheart Deal," *New York Times,* October 30, 2007.

232. Skoloff, "Everglades Restoration Has Almost Stopped."

233. National Research Council, *Progress toward Restoring the Everglades,* p. 10.

234. Quoted in Goodnough, "Efforts to Save Everglades Falter," p. 3.

235. Everglades superintendent Dan Kimball quoted in Cornelia Dean, "The Preservation Predicament," *New York Times,* January, 29, 2008, pp. D1, 4.

236. National Research Council, *Progress toward Restoring the Everglades,* p. 11.

237. Comments at Everglades Coalition Conference on Captiva Island, January 11, 2008.

238. Interview with author, January 7, 2008.

239. Interview with author, January 10, 2008.

240. Interview with author, January 7, 2008.

241. Damien Cave, "Florida Buying Big Sugar Tract for Everglades," *New York Times,* June 25, 2008.

242. Damien Cave, "A Dance of Environment and Economics in the Everglades," *New York Times,* July 31, 2008.

243. Damien Cave, "Everglades Plan under Fire as Vote Nears," *New York Times,* Decmeber 16, 2008, p. A25.

244. Damien Cave, "Harsh Review of Restoration in the Everglades," *New York Times,* September 30, 2008; National Research Council, *Progress toward Restoring the Everglades: The Second Biennial Review 2008* (Washington: National Academies Press, 2008).

245. National Academies of Sciences, "Effort to Restore Everglades Making Scant Progress." Press release, September 29, 2008, p. 1.

246. Interview with author, January 11, 2008.

247. Amy Leinbach Marquis, "Code Pink," *National Parks* 82, no. 3 (Summer 2008), pp. 24–25.

248. Tim Smight and Peter Frank, "10 Great Endangered Places to See While You Can," *USA Today,* April 18, 2008, p. 3D.

249. Ringo Starr, *Blast from Your Past* (New York: Capitol Records, 1975).

250. Douglas, *The Everglades: River of Grass,* p. 385.

Chapter Five

Epigraphs: Roosevelt quoted in E. A. Garrett, "The Grand Canyon: Are We Loving It to Death?" *National Geographic* 154, no. 1 (1978), p. 2. The National Park Service's Jan Balsom was interviewed by author March 7, 2008.

1. Michael P. Ghiglieri and Thomas M. Myers, *Death in Grand Canyon* (Flagstaff, Ariz.: Puma Press, 2001), p. 113.

2. U.S. Geological Survey, "Monitoring the Water Quality of the Nation's Large Rivers," USGS Fact Sheet FS-014-00 (Washington: 2000), p. 1.

3. Ibid.

4. Viscount Bryce, *Memories of Travel* (New York: Macmillan Company, 1923), p. 248.

5. Entry of August 13, 1869, in John Wesley Powell, *Report on Exploration of the Colorado River of the West and Its Tributaries* (Washington: U.S. Government Printing Office, 1875).

6. Marc Reisner, *Cadillac Desert* (New York: Penguin Books, 1987), p. 130.

7. Norris Hundley Jr., "The West against Itself," in *New Courses for the Colorado River,* edited by Gary Weatherford and F. Lee Brown (University of New Mexico Press, 1986), pp. 9–49; Helen Ingram, Lawrence Scaff, and Leslie Silko, "Replacing Confusion with Equity," in *New Courses for the Colorado River,* edited by Weatherford and Brown, pp. 177–199.

8. Helen Ingram, A. Dan Tarlock, and Cy R. Oggins, "The Law and Politics of the Operation of Glen Canyon Dam," in *Colorado River Ecology and Dam Management* (Washington: National Academy Press, 1991), p. 13; Scott K. Miller, "Undamming Glen Canyon: Lunacy, Rationality, or Prophecy?" *Stanford Environmental Law Journal* 19, no. 1 (2000), p. 131.

9. Reisner, *Cadillac Desert,* p. 136; Patrick McCully, *Silenced Rivers* (London: Zed Books, 1996); Donald Worster, *Rivers of Empire* (New York: Pantheon Books, 1985).

10. Philip Fradkin, *A River No More* (University of Arizona Press, 1981), p. 143; Miller, "Undamming Glen Canyon," p. 138.

11. Miller, "Undamming Glen Canyon," p. 152.

12. Roderick Frazier Nash, *Wilderness and the American Mind,* 4th ed. (Yale University Press, 2001), pp. 229–237; Reisner, *Cadillac Desert,* p. 295.

13. Colorado River Storage Project Act of 1956, 43 U.S. Code (U.S.C.) 620, Section 1.

14. Miller, "Undamming Glen Canyon," p. 160.

15. Calculated from data in U.S. Department of the Interior, *Budget Justifications F.Y. 1991* (Washington: GPO, 1991).

16. U.S. Bureau of Reclamation, *Operation of Glen Canyon Dam: Final Environmental Impact Statement* (Washington: U.S. Department of the Interior, 1995), pp. 114–115.

17. U.S. Bureau of Reclamation, "Glen Canyon Dam and Powerplant Self-Guided Tour" (Washington: GPO, 1998), p. 13.

18. E. A. Garrett, "The Grand Canyon: Are We Loving It to Death?" *National Geographic* 154, no. 1 (1978), p. 23.

19. U.S. Bureau of Reclamation, "Glen Canyon Dam and Powerplant Self-Guided Tour," p. 13.

20. U.S. General Accounting Office, "An Assessment of the Environmental Impact Statement on the Operations of the Glen Canyon Dam" (Washington: 1996), p. 197.

21. Ibid., p. 67.

22. Nash, *Wilderness and the American Mind,* p. 211.

23. Michael P. Ghiglieri, *Canyon* (University of Arizona Press, 1992), pp. 9, 190.

24. Ananda Shorey, "Canyon River Runner Numbers May Increase," *Arizona Republic,* December 24, 2004, p. D3; Candus Thomson, "Do-It-Yourself Rafters Want Bigger Share of River in Grand Canyon," *St. Louis Post-Dispatch,* October 13, 2002, p. T8. The NPS recently adopted a modified lottery system for granting private permits.

25. Edward Abbey, *The Monkey Wrench Gang* (Salt Lake City: Roaming the West, 1975).

26. Roderick Nash, "Wilderness Values and the Colorado River," in *New Courses for the Colorado River,* edited by Weatherford and Brown, pp. 211, 213.

27. Gilbert F. White, "A New Confluence in the Life of the River," in *New Courses for the Colorado River,* edited by Weatherford and Brown, pp. 222–23.

28. U.S. Bureau of Reclamation, *Operation of Glen Canyon Dam,* p. 3.

29. Richard Marks quoted in William MacDougall, "Will Grand Canyon Turn into a Lake of Mud?" *U.S. News and World Report,* September 28, 1981, p. 51.

30. Quoted in Philip Shabecoff, "Watt Drops a Power Project Threatening Colorado River," *New York Times,* October 30, 1981, p. A14.

31. Jeffrey W. Jacobs and James L. Wescoat Jr., "Managing River Resources: Lessons from Glen Canyon Dam," *Environment* 44, no. 2 (March 2002), p. 12; Miller, "Undamming Glen Canyon," p. 160.

32. U.S. Bureau of Reclamation, *Operation of Glen Canyon Dam,* p. 3.

33. Interview with author, December 21, 1998.

34. Ghiglieri, *Canyon*; William E. Schmidt, "Floods along Colorado River Set Off a Debate over Blame," *New York Times,* July 17, 1983, p. A1; Gary D. Weatherford. and F. Lee Brown, "Epilog," in *New Courses for the Colorado River,* edited by Weatherford and Brown, p. 226.

35. Ghiglieri and Myers, *Death in Grand Canyon,* p. 213.

36. Ibid., p. 2.

37. Doug George, "An Artificial Flood Does Good in the Grand Canyon," *Christian Science Monitor,* July 11, 2008, p. 25.

38. U.S. Bureau of Reclamation, *Operation of Glen Canyon Dam,* p. 2.

39. Ibid., p. 11; National Research Council, *River Resource Management in the Grand Canyon* (Washington: National Academies Press, 1996).

40. National Research Council, *River Resource Management in the Grand Canyon,* p. 17.

41. Ingram, Tarlock, and Oggins, "The Law and Politics of the Operation of Glen Canyon Dam," p. 10.

42. Jeffrey W. Jacobs and James L. Wescoat Jr., "Managing River Resources: Lessons from Glen Canyon Dam," *Environment* 44, no. 2 (March 2002), p. 13.

43. U.S. Congress, *Congressional Record,* November 27, 1991, p. S18742.

44. Grand Canyon Protection Act of 1992, Sec. 1802 (a), Public Law 102-575, Title XVIII.

45. Ibid., Sec. 1807.

46. Ibid., Sec. 1802 (b); U.S. GAO, "An Assessment of the Environmental Impact Statement on the Operations of the Glen Canyon Dam," p. 21.

47. U.S. Bureau of Reclamation, *Operation of Glen Canyon Dam,* p. 14.

48. Ibid., pp. 27–28.

49. Ibid., pp. 34–37.

50. Ibid., p. 53.

51. Richard J. Ingebretsen, "Foreword," *Stanford Environmental Law Journal* 19, no. 1 (2000), p. xi.

52. Glen Canyon Institute, "Citizens' Environmental Assessment on the Decommissioning of Glen Canyon Dam" (Salt Lake City: 2000), p. 4.

53. James Brooke, "In the Balance, the Future of a Lake," *New York Times,* September 22, 1997, p. A1; Miller, "Undamming Glen Canyon," p. 122.

54. Daniel P. Beard, "Dams Aren't Forever," *New York Times,* October 6, 1997.

55. David Brower, *The Place No One Knew,* with Eliot Porter (San Francisco: Sierra Club Books, 1963).

56. Edward Abbey, *Desert Solitaire* (New York: Ballantine Books, 1968), p. 174.

57. Powell, *Report on Exploration of the Colorado River,* July 31, p. 70; Wallace Stegner, *Beyond the Hundredth Meridian* (New York: Penguin Books, 1992, orig. published 1953), p. 90.

58. William H. Calvin, *The River That Flows Uphill* (San Francisco: Sierra Club Books, 1986), p. 71.

59. William Dolnick, *Down the Great Unknown* (New York: Harper Collins, 2002), p. 201.

60. Executive director Joan Nevills-Staveley, quoted in Patrick Graham, "Lake's Beneficiaries Start to Fret about Going Down Drain," *Salt Lake Tribune,* March 8, 1999, p. A1.

61. Mayor Michael A. Woods, quoted in James Brooke, "In the Balance, the Future of a Lake," *New York Times,* September 22, 1997, p. A1.

62. Michael E. Long, "The Grand Managed Canyon," *National Geographic* 192, no. 1 (July 1997), p. 134.

63. Abbey, *Desert Solitaire,* p. 173.

64. Worster, *Rivers of Empire,* p. 274.

65. Dolnick, *Down the Great Unknown,* p. 201.

66. Michael Satchell, "Power and the Glory," *U.S. News and World Report,* January 21, 1991, p. 70.

67. Andrew Murr and Sharon Begley, "Dams Are Not Forever," *Newsweek,* November 24, 1997, p. 48.

68. Miller, "Undamming Glen Canyon," p. 173; National Research Council, *River Resource Management in the Grand Canyon* , p. 1.

69. Ingram, Tarlock, and Oggins, "The Law and Politics of the Operation of Glen Canyon Dam.," p. 25.

70. Interview with author, December 23, 1999.

71. Eric Brazil, "Back to Nature?" *San Francisco Examiner,* August 24, 1997, pp. A1, A10.

72. Miller, "Undamming Glen Canyon," p. 179.

73. Dolnick, *Down the Great Unknown,* p. 201.

74. Reisner, *Cadillac Desert,* p. 492.

75. Glen Canyon Institute, "Citizens' Environmental Assessment on the Decommissioning of Glen Canyon Dam" (Salt Lake City: 2000); Miller, "Undamming Glen Canyon," p. 171.

76. Glen Canyon Institute, "Citizens' Environmental Assessment on the Decommissioning of Glen Canyon Dam."

77. Quoted in Michael Tennesen, "A River Runs through Them," *National Parks* (Winter 2006), p. 45.

78. Brooke, "In the Balance, the Future of a Lake," p. A1.

79. Beard, "Dams Aren't Forever"; Glen Canyon Institute, "Citizens' Environmental Assessment on the Decommissioning of Glen Canyon Dam."

80. Beard, "Dams Aren't Forever."

81. Quoted in Brooke, "In the Balance, the Future of a Lake" p. A1.

82. Ingebretsen, "Foreword," *Stanford Environmental Law Journal,* p. xiv.

83. Larry Tarp quoted in Brooke, "In the Balance, the Future of a Lake" p. A1.

84. Jason Zengerle, "Water over the Dam," *New Republic,* November 24, 1996, p. 21.

85. Mark Muro, "Can the River Run Again?" *Arizona Daily Star,* April 20, 1997, p. 1E.

86. Quoted in Brooke, "In the Balance, the Future of a Lake" p. A1.

87. Zengerle, "Water over the Dam," p. 21.

88. Interview with author, November 22, 2000.

89. Brooke, "In the Balance, the Future of a Lake," p. A1.

90. Murr and Begley, "Dams Are Not Forever," p. 48.

91. Subcommittee on National Parks and Public Lands and the Subcommittee on Water and Power of the House Committee on Resources, *Joint Hearing on the Sierra Club's Proposal to Drain Lake Powell or Reduce Its Water Storage Capacity,* 105th Cong., 1st sess., September 24, 1997, p. 11.

92. Ibid.

93. Brazil, "Back to Nature," p. A10.

94. Beard, "Dams Aren't Forever."

95. U.S. Bureau of Reclamation, *Operation of Glen Canyon Dam,* p. 168; Miller, "Undamming Glen Canyon," p. 185; National Research Council, *River Resource Management in the Grand Canyon,* p. 167.

96. National Research Council, *River Resource Management in the Grand Canyon,* p. 1.

97. Ingebretsen, "Foreword," p. xiii.

98. Beard, "Dams Aren't Forever," p. B6.

99. Muro, "Can the River Run Again?" p. 1E.

100. Miller, "Undamming Glen Canyon," p. 188; National Research Council, *River Resource Management in the Grand Canyon,* p. 171.

101. Brazil, "Back to Nature," p. A10; Miller, "Undamming Glen Canyon," p. 190.

102. Brooke, "In the Balance," p. A1.

103. National Research Council, *River Resource Management in the Grand Canyon,* p. 120.

104. Brazil, "Back to Nature," p. A10; Muro, "Can the River Run Again?" p. 1E.

105. Miller, "Undamming Glen Canyon," p. 175.

106. Brooke, "In the Balance," p. A1; Zengerle, "Water over the Dam," p. 22.

107. Miller, "Undamming Glen Canyon," p. 175.

108. Ibid., p. 191; Zengerle, "Water over the Dam," p. 22.

109. Glen Canyon Institute, "Citizens' Environmental Assessment on the Decommissioning of Glen Canyon Dam," p. 2.

110. Miller, "Undamming Glen Canyon," p. 171.

111. William R. Lowry, *Dam Politics* (Georgetown University Press, 2003); Miller, "Undamming Glen Canyon," pp. 164–167.

112. U.S. Bureau of Reclamation, *Operation of Glen Canyon Dam,* p. 11; National Research Council, *River Resource Management in the Grand Canyon.*

113. National Research Council, *River Resource Management in the Grand Canyon,* p. 71.

114. Ingebretsen, "Foreword," p. xiii.

115. U.S. Bureau of Reclamation, *Operation of Glen Canyon Dam,* p. 5.

116. U.S. GAO, "An Assessment of the Environmental Impact Statement on the Operations of the Glen Canyon Dam," p. 67.

117. U.S. Fish and Wildlife Service, *Final Biological Opinion: Operation of Glen Canyon Dam* (Washington: 1994); U.S. GAO, "An Assessment of the Environmental Impact Statement on the Operations of the Glen Canyon Dam."

118. U.S. GAO, "An Assessment of the Environmental Impact Statement on the Operations of the Glen Canyon Dam," p. 89.

119. Ibid., p. 95.

120. Bureau statement in ibid., p. 97.

121. Interview with author, November 25, 2008.

122. Tom Moody, "Glen Canyon Dam: Coming to an Informed Decision," *Grand Canyon Trust* (Fall 1997), p. 2.

123. Interview with author, February 8, 2007.

124. Lowry, *Dam Politics,* chapter 4.

125. Alexandra Ravinet, "Rivers Get over the Dam," *Christian Science Monitor,* July 8, 1999, p. 14.

126. Lowry, *Dam Politics,* p. 77.

127. Murr and Begley, "Dams Are Not Forever," p. 48.

128. Brad Udall, "Arizona at a Crossroads over Water and Growth," *Arizona Republic,* March 9, 2008.

129. Tim Folger, "Requiem for a River," *Onearth* (Spring 2008), pp. 24–35.

130. Jeanne Nienaber Clarke and Daniel C. McCool, *Staking out the Terrain* (State University of New York Press, 1996); Lowry, *Dam Politics,* pp. 30–31.

131. Interview with Norm Henderson, November 24, 1998.

132. Interview with author, March 7, 2008.

133. Beard, "Dams Aren't Forever."

134. Murr and Begley, "Dams Are Not Forever," p. 48.

135. Interview with author, December 21, 1998.

136. Ingebretsen, "Foreword," p. xiv; Miller, "Undamming Glen Canyon," p. 206.

137. Long, "The Grand Managed Canyon," p. 130.

138. Jacobs and Wescoat, "Managing River Resources: Lessons from Glen Canyon Dam," p. 17.

139. National Research Council, *Downstream: Adaptive Management of Glen Canyon Dam and the Colorado River Ecosystem* (Washington: National Academies Press, 1999), p. 2.

140. U.S. Bureau of Reclamation, *Operation of Glen Canyon Dam,* p. 22.

141. Ibid., p. 29.

142. Long, "The Grand Managed Canyon," p. 126.

143. U.S. Bureau of Reclamation, *Operation of Glen Canyon Dam,* p. 41.

144. Record of Decision (www.gcmre.gw/amwg_new/Documents/RTC/Trc 12-16.htm).

145. Interview with author, November 24, 1998.

146. L. H. Gunderson, C. S. Holling, and Stephen S. Light, *Barriers and Bridges to the Renewal of Ecosystems and Institutions* (Columbia University Press, 1995); C. J. Walters, L. Gunderson, and C. S. Holling, "Experimental Policies for Water Management in the Everglades," *Ecological Applications* 2 (1992), pp. 189–202.

147. National Research Council, *Downstream: Adaptive Management of Glen Canyon Dam,* p. 55.

148. Ibid., p. 133.

149. Ibid., p. 17.

150. Interview with author, November 24, 1998.

151. National Research Council, *Downstream: Adaptive Management of Glen Canyon Dam,* p. 7.

152. Ibid., p. 9.

153. Interview with author, March 8, 2002.

154. Jacobs and Wescoat, "Managing River Resources," p. 18.

155. Interview with author, April 13, 2001.

156. Michael P. Collier, Robert H. Webb, and Edmund D. Andrews, "Experimental Flooding in Grand Canyon," *Scientific American* (January 1997), p. 85; A. D. Konieczki, Julia B. Graf, and Michael C. Carpenter, "Streamflow and Sediment Data," report for the U.S. Geological Survey 97-224 (Washington: USGS, 1997).

157. Associated Press, "Flood Is Called Right Tonic for Grand Canyon," *New York Times,* April 14, 1996, p. A32.

158. Collier, Webb, and Andrews, "Experimental Flooding in Grand Canyon," p. 87; Konieczki, Graf, and Carpenter, "Streamflow and Sediment Data."

159. Quoted in William K. Stevens, "A Dam Open, Grand Canyon Roars Again," *New York Times,* February 25, 1997, pp. B7, B12.

160. Quoted in Brazil, "Back to Nature," p. A10.

161. Interview with author, November 23, 1998.

162. Quoted in Brent Israelson, "Artificial Flood Failed to Solve Grand Canyon Problems, Scientists Say," *Salt Lake Tribune,* April 10, 2002.

163. Sandra Blakeslee, "Restoring an Ecosystem Torn Asunder by a Dam," *New York Times,* June 11, 2002, p. F1.

164. Associated Press, "Flood Is Called Right Tonic for Grand Canyon"; Collier, Webb, and Andrews, "Experimental Flooding in Grand Canyon," p. 89; Long, "The Grand Managed Canyon."

165. Miller, "Undamming Glen Canyon," p. 162.

166. Sandra Blakeslee, "U.S. Panel Backs a Risky Effort to Save a Grand Canyon Fish," *New York Times,* April 26, 2002, p. A16; Stevens, "A Dam Open."

167. Interview with author, January 25, 2001.

168. Barry D. Gold, "Preliminary Results of Low Steady Summer Flows Test," unpublished manuscript, Grand Canyon Monitoring and Research Center, 2000; U.S. Geological Survey, "Colorado River/Grand Canyon Focus of Science Meeting," press release, April 23, 2001.

169. Israelson, "Artificial Flood Failed to Solve Grand Canyon Problems, Scientists Say."

170. Jim Carlton, "Environmentalists, Power Authorities Fight over Grand Canyon Flood Plan," *Wall Street Journal,* April 17, 2002, p. 1; Anne Minard, "Another Canyon Flood?" *Arizona Daily Sun,* April 18, 2002, p. A1.

171. Blakeslee, "Restoring an Ecosystem Torn Asunder by a Dam," p. F1; Blakeslee, "U.S. Panel Backs a Risky Effort to Save a Grand Canyon Fish."

172. Sandra Blakeslee, "In Bold Experiment at Canyon, a River Rips through It," *New York Times,* November 23, 2004, p. F4; U.S. Geological Survey, "Interior Scientists to Evaluate Effects of High Flow Test at Glen Canyon Dam," press release, November 19, 2004.

173. Associated Press, "Fish Numbers Drop after Grand Canyon Flood," MSNBC.com, March 8, 2005; Patrick O'Driscoll, "River 'Flush' Washes Up New Data," *USA Today,* October 27, 2005, p. A3.

174. William R. Lowry, *Preserving Public Lands for the Future* (Georgetown University Press, 1998), p. 92; National Research Council, *Haze in the Grand Canyon* (Washington: National Academies Press, 1990).

175. Lowry, *Preserving Public Lands for the Future,* pp. 91–92; U.S. National Park Service, "Report on Effects of Aircraft Overflights on the National Park System" (Washington: 1994); interview with Jan Balsom, March 7, 2008.

176. Peter Jacques and David M. Ostergren, "The End of Wilderness: Conflict and Defeat of Wilderness in the Grand Canyon," *Review of Policy*

Research 23, no. 2 (2006), pp. 581–82; Lowry, *Preserving Public Lands for the Future,* p. 92; Bo Shelby and Joyce M. Nielson, "Motors and Oars in the Grand Canayon," report prepared for the National Park Service (Washington: U.S. NPS, 1976).

177. Interview with Jan Balsom, March 7, 2008.

178. Scott Kirkwood, "A Radioactive Proposition," *National Parks* 82, no. 4 (2008), pp. 10–11.

179. Todd Wilkinson, "Homecoming," *National Parks* (May 1996), pp. 40–45.

180. Juliet Eilperin, "Effort to Renew Colorado River Launched," *Washington Post,* September 15, 2004, p. A03; Miller, "Undamming Glen Canyon," p. 163.

181. Interview with Sam Spiller, March 11, 2008.

182. Quotes from Eilperin, "Effort to Renew Colorado River Launched."

183. Jeffrey E. Lovich and Theodore S. Melis, "Lessons from 10 Years of Adaptive Management in Grand Canyon," in *The State of the Colorado River Ecosystem in Grand Canyon,* edited by S. P. Gloss, Jeffrey E. Lovich, and Theodore S. Melis (Reston, Va.: U.S. Geological Survey, 2005), p. 207.

184. Ibid., p. 208.

185. Chris Ayres, "Grand Canyon Is Flooded to Save a Rare Fish," *Times of London,* March 7, 2008, p. 46; Janet Wilson, "Feds' Plan to 'Flush' Grand Canyon Stirs Concerns," *Los Angeles Times,* March 4, 2008.

186. Quoted in Israelson, "Artificial Flood Failed to Solve Grand Canyon Problems, Scientists Say."

187. Quoted in Associated Press, "Fish Numbers Drop after Grand Canyon Flood," MSNBC.com, March 8, 2005.

188. Interview with author, February 9, 2007.

189. Interview with author, March 13, 2007.

190. U.S. Bureau of Reclamation, "Long-Term Experimental Plan for Operation of Glen Canyon Dam to Be Developed," press release, November 3, 2006.

191. Interview with author, March 13, 2007.

192. Joe Baird, "Lake Powell May Never Be Full Again," *Salt Lake Tribune,* April 2, 2006.

193. Randal C. Archibold, "Western States Agree to Water-Sharing Pact," *New York Times,* December 10, 2007; Patrick O'Driscoll, "Seven States Sign Colorado River Water Pact," *USA Today,* December 14, 2007, p. A3.

194. Felicity Barringer, "Lake Mead Could Be within a Few Years of Going Dry, Study Finds," *New York Times,* February 13, 2008.

195. Frank Clifford, "Troubled Waters," *Los Angeles Times,* May 25, 2008, p. A1.

196. Interview with Sam Spiller, November 25, 2008.

197. Tennesen, "A River Runs through Them," p. 45.

198. Interview with Barry Gold, March 21, 2008.

199. Quoted in Associated Press, "Man-Made Flood Rushes through Grand Canyon," FoxNews.com, March 5, 2008.

200. George, "An Artificial Flood Does Good in the Grand Canyon," p. 25.

201. Ibid.

202. Quoted in Dan Glaister, "Environmentalists and Park Workers Criticize Grand Canyon Flooding," Guardian.co.uk, March 4, 2008.

203. Wilson, "Feds' Plan to 'Flush' Grand Canyon Stirs Concerns."

204. Interview with author, March 7, 2008.

205. Interview with author, March 11, 2008.

206. Wilson, "Feds' Plan to 'Flush' Grand Canyon Stirs Concerns."

207. Public Employees for Environmental Responsibility, "Interior Department Stages Grand Canyon Green Wash," press release, March 3, 2008.

208. Quoted in Associated Press, "Man-Made Flood Rushes through Grand Canyon."

209. Interview with author, March 7, 2008.

210. George, "An Artificial Flood Does Good in the Grand Canyon."

211. Lovich and Melis, "Lessons from 10 Years of Adaptive Management in Grand Canyon," p. 218.

212. Interview with author, March 11, 2008.

213. Interview with author, November 25, 2008.

214. Interview with author, March 7, 2008.

215. Interview of March 21, 2008.

216. Interview with author, March 11, 2008.

217. Interview with author, March 7, 2008.

218. Bruce Springsteen, *The Rising* (New York: Columbia Records, 2002).

Chapter Six

Epigraphs: Brian Jones and Frank Baumgartner from their edited volume, *Policy Dynamics* (University of Chicago Press, 2002), p. 306. Yellowstone superintendent Mike Finley was interviewed by the author, July 15, 1996.

1. Associated Press, "Artificial Flood Created to Rejuvenate the Grand Canyon," *New York Times,* March 27, 1996, p. B8.

2. See the foreword by William Cronon in David Louter's *Windshield Wilderness* (University of Washington Press, 2006), pp. ix–xii.

3. See also Louter, *Windshield Wilderness,* on automobile use in Washington's national parks.

4. For more, see William R. Lowry, *Dam Politics* (Georgetown University Press, 2003), chapter 6.

5. National Park Service, "Secretary Salazar Announces $750 Million Investment," NPS news release, April 22, 2009.

6. Associated Press, "Florida: Stimulus for Everglades," *New York Times,* April 30, 2009, p. A17.

7. Quoted in Michael Grunwald, *The Swamp* (New York: Simon & Schuster, 2007), p. 353.

8. Leslie Bella, *Parks for Profit* (Montreal: Harvest House, 1987); William R. Lowry, *Preserving Public Lands for the Future* (Georgetown University Press, 1998).

9. For an example, from the front page of the *Globe and Mail,* see Alana Mitchell, "Banff's Outlook Not a Pretty Picture," *Globe and Mail,* December 24, 1994, p. A1.

10. Banff–Bow Valley Task Force, *Banff–Bow Valley: At the Crossroads* (Ottawa: Minister of Supply and Services Canada, 1996).

11. William R. Lowry, "Can Bureaucracies Change Policy?" *Journal of Policy History* 20, no. 2 (2008), pp. 287–306.

12. Judith A. Layzer, *The Environmental Case* (Washington: CQ Press, 2006), chapter 11; Barry G. Rabe, *Statehouse and Greenhouse* (Brookings, 2004).

13. Rabe, *Statehouse and Greenhouse,* pp. 30–31.

14. Ibid., pp. 19-20.

15. Frank R. Baumgartner and Bryan D. Jones, *Agendas and Instability in American Politics* (University of Chicago Press, 1993), chapter 4; Robert J. Duffy, *Nuclear Politics in America* (University Press of Kansas, 1997).

16. Baumgartner and Jones, *Agendas and Instability in American Politics*; M. Burnett and C. Davis, "Getting Out the Cut," *Administration and Society* 34, no. 2 (2002), pp. 202–28; Sarah B. Pralle, "Venue Shopping, Political Strategy, and Policy Change," *Journal of Public Policy* 23, no. 3 (2003), pp. 233–260.

17. Baumgartner and Jones, *Agendas and Instability in American Politics*; Bryan D. Jones and Frank R. Baumgartner, *The Politics of Attention* (University of Chicago Press, 2005).

18. Daniel Kahneman and Amos Tversky, "Choices, Values, and Frames," *American Psychologist* 39 (1984), p. 343.

19. Ibid., p. 348.

20. William P. Bottom and Amy Studt, "Framing Effects and the Distributive Aspect of Integrative Bargaining," *Organizational Behavior and Human Decision Processes* 56 (1993), p. 459.

21. William P. Bottom, "Negotiator Risk," *Organizational Behavior and Human Decision Processes* 76, no. 2 (November 1998), pp. 89–112; Rose McDermott, *Risk-Taking in International Politics* (University of Michigan Press, 1998), p. 4; Barry O'Neill, "Risk Aversion in International Relations Theory," *International Studies Quarterly* 45, no. 4 (2001), pp. 622, 625.

22. Jack S. Levy, "Prospect Theory, Rational Choice, and International Relations," *International Studies Quarterly* 41, no. 1 (1997), pp. 87–112; O'Neill, "Risk Aversion in International Relations Theory," p. 621.

23. James N. Druckman and Kjersten R. Nelson, "Framing and Deliberation," *American Journal of Political Science* 47, no. 4 (2003), pp. 729–45; William G. Jacoby, "Issue Framing and Public Opinion on Government Spending," *American Journal of Political Science* 44, no. 4 (2000), pp. 750–67; Mark R. Joslyn, "Framing the Lewinsky Affair: Third-Person Judgments by Scandal Frame," *Political Psychology* 24, no. 4 (2003), pp. 829–44.

24. Matthew D. Adler and Eric A. Posner, *New Foundations of Cost-Benefit Analysis* (Harvard University Press, 2006), pp. 3–4; K. W. Easter, N. Becker, and S. O. Archibald, "Benefit-Cost Analysis and Its Use in Regulatory Decisions," in *Better Environmental Decisions,* edited by K. Sexton and others (Washington: Island Press, 1999), pp. 160–61.

25. Scott K. Miller, "Undamming Glen Canyon: Lunacy, Rationality, or Prophecy?" *Stanford Environmental Law Journal* 19, no. 1 (2000), pp. 121–207.

26. Adler and Posner, *New Foundations of Cost-Benefit Analysis,* p. 121; Daniel A. Mazmanian and Jeanne Nienaber, *Can Organizations Change?* (Brookings Institution, 1979), pp. 22–23.

27. H. B. Leonard and R. J. Zeckhauser, "Cost-Benefit Analysis Defended," in *Environmental Ethics,* edited by D. Schmidtz and E. Willott (Oxford University Press, 2002), p. 465.

28. Mazmanian and Nienaber, *Can Organizations Change?* pp. 22–23.

29. Easter, Becker, and Archibald, "Benefit-Cost Analysis," p. 172.

30. Steven Kelman, "Cost-Benefit Analysis: An Ethical Critique," in *Environmental Ethics,* edited by Schmidtz and Willott, pp. 455-62; Andrew Brennan, "Moral Pluralism and the Environment," in *Environmental Ethics.*

31. Easter, Becker, and Archibald, "Benefit-Cost Analysis," p. 167.

32. Brennan, "Moral Pluralism and the Environment," p. 467; Mark Sagoff, *The Economy of the Earth* (Cambridge University Press, 1988).

33. Frank Ackerman and Lisa Heinzerling, *Priceless: On Knowing the Price of Everything and the Value of Nothing* (New York: New Press, 2004), pp. 39–40; Kelman, "Cost-Benefit Analysis," p. 461.

34. Adler and Posner, *New Foundations of Cost-Benefit Analyis,* p. 159.

35. Leonard and Zeckhauser, "Cost-Benefit Analysis Defended," p. 464.

36. David Schmidtz, "A Place for Cost-Benefit Analysis," in *Environmental Ethics,* edited by Schmidtz and Willott, p. 485.

37. Adler and Posner, *New Foundations of Cost-Benefit Analysis,* p. 189; Easter, Becker, and Archibald, "Benefit-Cost Analysis," p. 161.

38. L. H. Gunderson, C. S. Holling, and Stephen S. Light, *Barriers and Bridges to the Renewal of Ecosystems and Institutions* (Columbia University Press, 1995); K. N. Lee, "Appraising Adaptive Management," *Conservation Ecology* 3, no. 2 (1999), p. 3.

39. C. J. Walters, L. Gunderson, and C. S. Holling, "Experimental Policies for Water Management in the Everglades," *Ecological Applications* 2 (1992),

pp. 189–202; U.S. Government Accountability Office, *South Florida Ecosystem Restoration: Task Force Needs to Improve Science Coordination to Increase the Likelihood for Success* (Washington: 2003); National Research Council, *Adaptive Management for Water Resources Project Planning* (Washington: National Academies Press, 2004).

40. National Research Council, *Adaptive Management,* p. 58.

41. William C. Clark, "Adaptive Management, Heal Thyself," *Environment* 44, no. 2 (March 2002), p. 7.

42. Ibid., p. 7.

43. Jeffrey L. Pressman and Aaron Wildavsky, *Implementation* (University of California Press, 1973); Paul A. Sabatier and Daniel Mazmanian, "The Implementation of Public Policy: A Framework for Analysis," *Policy Studies Journal* 8 (1980), pp. 538–60.

44. Paul R. Schulman, "Nonincremental Policy Making," *American Political Science Review* 69 (1975), p. 1367.

45. Interview with author, January 7, 2008.

46. For an early example, see Theodore J. Lowi, "American Business, Public Policy, Case Studies, and Political Theory," *World Politics* 16 (1964), pp. 677–715.

47. Jeanne Nienaber Clarke and Daniel C. McCool, *Staking Out the Terrain* (State University of New York Press, 1996); Pressman and Wildavsky, *Implementation.*

48. Interview with author, July, 17, 2008.

INDEX